NUMERICAL CONTROL

OXFORD SERIES ON ADVANCED MANUFACTURING

SERIES EDITORS

J. R. CROOKALL
MILTON C. SHAW
NAM P. SUH

1. William T. Harris. *Chemical Milling: The Technology of Cutting Materials by Etching* (1976)
2. Bernard Crossland. *Explosive Welding of Metals and its Applications* (1982)
3. Milton C. Shaw. *Metal Cutting Principles* (1984)
4. Shiro Kobayashi, Soo-Ik Oh, Taylan Altan. *Metal Forming and the Finite-Element Method* (1989)
5. E. Taniguchi. *Energy Beam Processing of Materials* (1989)
6. Nam P. Suh. *The Principles of Design* (1990)
7. N. Logothetis and H. P. Wynn. *Quality through Design* (1990)
8. John L. Burbidge. *Production Flow Analysis* (1990)
9. J. Francis Reintjes. *Numerical Control: Making a New Technology* (1991)

NUMERICAL CONTROL

Making a New Technology

J. FRANCIS REINTJES

Professor Emeritus and Senior Lecturer, Electrical Engineering

Massachusetts Institute of Technology

New York Oxford

OXFORD UNIVERSITY PRESS

1991

Oxford University Press

Oxford New York Toronto
Delhi Bombay Calcutta Madras Karachi
Petaling Jaya Singapore Hong Kong Tokyo
Nairobi Dar es Salaam Cape Town
Melbourne Auckland

and associated companies in
Berlin Ibadan

Published by Oxford University Press, Inc.,
200 Madison Avenue, New York, New York 10016

Library of Congress Cataloging-in-Publication Data
Reintjes, J. Francis.
Numerical control: making a new technology / J. Francis Reintjes.
p. cm.—(Oxford series on advanced manufacturing)
Includes bibliographical references and index.
ISBN 0-19-506772-X
1. Machine-tools—Numerical control—History. I. Title. II. Series.
TJ1189.R37 1991
621.9'023—dc20 90-43298

2 4 6 8 9 7 5 3 1

Printed in the United States of America
on acid-free paper

FOREWORD

Numerical Control: Making a New Technology will be of great benefit to readers whose principal interest is to learn about the ups and downs in bringing a particular new technology from concept in a laboratory to the production shop floor. Often we fail to realize that the acceptance of a new technology or machine by any society, and especially ours, involves not only the properties of the machine, but the nature of the "societal software"—the doctrine and goals, the regulations, the political, the economic, the social and the educational systems—in effect the total culture which gives life, vitality and stability to the society in which we live. For these reasons a subtitle such as *Guidelines for Achieving viable Relationships between Government, Industry and Technical Universities to Enhance a Nation's Industrial Strength* would indicate additional important contributions of the book.

History shows that it has often taken twenty to forty years to achieve widespread acceptance of even the most momentous of new technologies—witness the time span between the invention of the jet engine and its commercial use, or between the invention of the transistor and its use in satellite communications or the personal digital computer. During those time spans, vast changes occurred in our way of life, and as each change occurred, the new technology was exploited for its ability to contribute to the change. Thus, the demonstration of a full-scale pilot model of a numerically controlled milling machine was a mere prelude to the total effort expended in getting the concept of real-time digital control applied to a broad array of processes. Today, digital computers are the central feedback control elements in such diverse applications as chemical process control systems, spacecraft guidance systems and automobile engine-control systems.

The Reintjes book becomes available during a period when the United States is experiencing a decline from the enviable position it held shortly after World War II as the world leader in manufacturing productivity and global competitiveness. Until about two decades ago, it was beyond the concern of many in government, management, and labor that such a decline could ever exist. Until then, American products were in wide demand and sold everywhere. The future looked bright but something happened along the way. In the view of many this decline is among the principal causes of our unfavorable balance of trade, our huge national debt, and much instability in the labor market. The trade imbalance and the national debt are of concern to our political and industrial leaders, and also to many of our trading partners. The instability in the labor market has caused great stress among large sectors of the work force where even highly skilled workers have become unemployed, often unemployable, as older industries have withered away. Many workers are called upon to learn new skills late in life for which they are unprepared, often the result of being deprived of knowledge and attitude toward continued learning that should have been given them in their youth, or because their adaptability to learn anew is greatly lessened as they have approached or passed middle age.

Underlying many of these matters is the simple fact that we live in a world of great change and that we, as individuals, are the principal instigators of the change. As we act to improve our lot and achieve more abundance, we increase our expectations. Then, by a process of feedback into our social system, these expanding expectations cause more change (1). An assessment of history shows that ever since the dawn of man no machine or machine system has been invented to serve mankind which, within a relatively short while, has not become inadequate. Society continually demands the new to replace the old—the horse and buggy were displaced by the automobile, for many tasks the railroad has been displaced by the propeller-driven airplane which in turn has yielded to the supersonic jet. The consequence is perpetual change which all too often catches the average citizen unprepared. No longer can anyone assume that merely by graduating from high school, or even a university, he or she will know enough to last for the next fifty years.

Different authorities have advanced numerous reasons for our present condition. Some cite intense international competition; some fault the attitudes of management; others believe our national investment in research and development is inadequate; and still others blame outworn doctrines of labor. Many fault our education system, not only in the elementary and high schools, but also in the universities and on-the-job in industry. Unfortunately no simple solutions exist, certainly no single solution, because of the interconnectedness of things in today's society—everything appears to affect everything else. Stated briefly, the world has experienced an avalanche of new knowledge and a host of new technologies. We face a dynamic-system problem whose understanding demands new ways of exploring and evaluating situations, and whose solution calls for new policies and procedures by government, industry, and education. All must consider a wider range of issues, adopt a longer term view, and draw upon new talents and specialties.

I conclude that much of the debate about the causes of our industrial decline is misdirected and leads people to believe that the problem is rooted primarily in technology and can be corrected merely by upgrading or fine-tuning the kinds of machines we have used for most of the industrial revolution. But these machines have been upgraded continually throughout the past century and for much of that time the valuable property for industry to own was "great skill at fine-tuning a craft." There was widespread belief that the skills a worker learned in his youth would last a lifetime, and the rate that jobs were upgraded would not displace a large number of employees. Now we know this doctrine is false. We are faced with wholly new technologies built on knowledge, largely science based, that did not exist even a decade or so ago. As a consequence, the most valuable properties to own now are: (1) the mastery of the fundamentals of modern science; (2) the technology for the processing of information; (3) the means to couple both for measurement and control of quality and productivity throughout business and industry; and (4) the skill to do these things in benign ways to improve our quality of life. Instead of merely an upgraded industrial revolution we face an *intellectual revolution*, which is something entirely new.

Central to our mastery of this new revolution is our ability to enhance the education of our citizens at all levels. The task is not a trivial one as we frequently encounter a shortage not only of qualified teachers, but of workers who are willing or even able to be taught. As stated recently by Mr C. Robert Wieser, a colleague of mine at the Massachusetts Institute of Technology during the period when the Whirlwind Computer was being developed: "It is one thing to explain something that lies only slightly beyond a person's previous experience, and yet another thing to explain something that lies outside his imagination (2)."

As awareness of these matters came to the attention of our national leadership, the National Academy of Engineering established a panel on Engineering Research Centers. The report of the panel released in 1984 set forth guidelines for structuring a number of centers to be established on selected university campuses (3). Funds were to be furnished by the Federal Government through the National Science Foundation (NSF). By the Fall of 1986 the NSF had established 11 centers; there are now 19. The panel identified the mission of the centers "to conduct cross-disciplinary on-campus programs that will forge new alliances between industry and academia in engineering education." This statement undoubtedly was based on the belief that real-world problems are inherently interdisciplinary, and are not handled properly by a narrowly focused disciplinary group. The discovery of much new knowledge is inevitably one result of the research, which then must be taught in the appropriate form to many others. A university campus is a natural environment in which this teaching can happen quickly, but in industrial plants it seldom and often never happens.

As NSF university research centers become established and it becomes possible for graduate students in engineering to play a major role, and where new teaching programs go hand in hand with research, it will be possible for our nation to accomplish several priceless benefits for the price of one. First, it will achieve high quality research with an imaginative forward-looking vision and through the teaching process, the new knowledge discovered in the laboratories is first taught to graduate students, where it becomes so well understood that it soon is taught to undergraduates. This is no mere trickle-down process but is rather a natural result of an imaginative professor desiring to work toward achieving his vision of the future by teaching his students what he believes they will need to know a decade or so hence. Thereby he will maximize the probability that his vision will become reality. Second, a research laboratory located on a university campus inherently operates in a manner that tends to ensure that it will remain viable and alive. There are many instances where it has been observed that industrial or government research laboratories flourish for about their first decade, or while they are growing, but when there is no infusion of new blood or turnover of staff, they appear to atrophy. However, a university-based research laboratory staffed by an interdisciplinary group of faculty and graduate students has almost guaranteed viability because as its student-complement of staff finish their degree programs and move on, new blood is recruited to take their place. The staff who leave will possess the most up-to-date knowledge of their time, they will know where the frontiers are and can work to push them forward. Many will innovate and create wholly new enterprises.

But when establishing strategies for the participation of government and industry with a technical university, it is essential to realize that the environment, the responsibilities, and the goals of a university are very different from those of either government or industry. Sir Eric Ashby (4) in his essay on universities and the technologic revolution captures the essence of the relationships between the various cultures. Only in recent times have many universities adapted to the mores of scientific thought. The same cannot be said for the mores of technology. As Ashby says:

> There is no great divergence between the attitude of the physicist toward the concept of entropy and the attitude of the philosopher toward the concept of virtue. But teaching and research in technology are unashamedly tendentious, and their tendentiousness has not been mellowed (as it has for medicine and law) by centuries of tradition. Technology is of the earth, earthy; it is susceptible to pressure from industry and government departments; it is under an obligation to deliver the goods. And so the crude engineer, the mere technologist (the mere adjectives are symptoms of the attitude) are tolerated in universities because the state and industry are willing to finance them (5).

Admittedly much of Ashby's essay is addressed to the British university system but the gist of his remarks pertain to the American system because, as our universities have become more tightly coupled to their Federal or State governments through subsidies, research grants, scholarships, and tuition aid and loans, they have acquired, in Ashby's words: "a new sort of patron . . . [the taxpayer]. It is no longer a bishop or a queen or merchant-prince, but the little man on the street who is the patron of learning (6)." In Ashby's view:

> He [the taxpayer] is liable to exercise what de Tocqueville called the despotism of the majority . . . The way to ensure autonomy without independence is to secure the confidence of patrons, and the way to secure this confidence is for the university to show a strong sense of responsibility to society. This is not easy, for the university's function is to give society not what it wants but what it

> needs; the two do not always coincide . . . To do this requires delicate diplomacy. If universities were to become community service stations responding to this and that popular demand, their integrity would be endangered and their influence weakened . . . Universities must of necessity be committed to society but the commitment should be on their own terms (7).

It is hoped that this volume can be used to the advantage of today's policymakers in industry, government and education as they respond to the demands of the industrial world of today and tomorrow. The stakes are enormous and the time scale entails looking ten to twenty years ahead. The work of the Servomechanisms Laboratory has shown that a research center in engineering located on a university campus possesses great promise for contributing to this end, provided its administration can nourish a strong interdisciplinary environment wherein information flows freely across departmental lines, from science to engineering to management to the humanities and to industry. For engineering students in particular, a project-type organization which permits students to participate as interns in new authentic engineering problems and take their share of responsibility for the quality of the work they perform under the guidance of a master, has many attractive features. Hopefully, as they enter the industrial scene as practicing engineers they will accept the doctrine of continued learning throughout their lifetimes and teach this doctrine to others.

Institute Professor Emeritus
Massachusetts Institute of Technology
Cambridge, MA
November 1990

Gordon S. Brown

1. Brown, G.S. "Engineering and Societal 'Software'—a New Imperative." *Technology Review* **75** (3): 2–11, 1973.
2. Forrester, J.W., R.D. Everett, C.R. Wieser. "Whirlwind's Genesis and Descendents," *The Computer Museum Report (Boston)*, p 16, Spring 1985.
3. Compton, W.D. Dale. "The Engineering Centers: Are We Prepared For The Challenge?" *The Bridge* (National Academy of Engineering), Spring 1987.
4. Ashby, E. *Technology and Academics.* New York: Macmillan, 1958.
5. Ibid p 66.
6. Ibid p 89.
7. Ibid pp 90–94, *passim.*

PREFACE

The engineering research program described in this book stemmed from a national need perceived by the US Air Force during the late 1940s and early 1950s for machine tool systems that could manufacture the complex parts required for production of future high-speed, high-performance aircraft. At the time the program began, very little was known about servomechanisms that operated from pulse type input signals, and knowledge of real-time digital control of processes was essentially nonexistent. The project addressed these matters in the context of real-time digital control of machine tools, and from the engineering research effort there emerged a three-axis digitally controlled continuous-path milling machine which served as a testbed for study and evaluation of its merits and as a model for others to build upon. The new technology became known as numerical control, or simply NC.

Oftentimes one technological innovation requires another in order to make the first one a practical reality. This was so in the case of numerical control. A major need was to simplify and quicken the writing of computer programs for NC machines in order to put the new technology on a firm economic footing. The Laboratory's response to this need was the development of a powerful special purpose programming system which became known as the APT system, an acronym for automatically programmed tool system. Unfortunately computer programs and programming systems were not patentable under the patent laws of those days; had a patent been possible, APT most certainly would have qualified as a candidate invention in light of its revolutionary concepts.

The third project of the overall program was a natural outgrowth of the other two; it addressed the broader question of how digital computers could serve in manufacturing processes as a whole—particularly during the design stage. Although emphasis was on the design of mechanical parts for aircraft, computer-aided electronic circuit design and electrical network design were also included in our work.

As stated in Chapter 1, several factors prompted the preparation of the text: A desire to record events that transpired from the viewpoint of one who had a hand in many of the important decisions made during the evolution of the technology; to give insights into a piece of university research that has been judged highly successful in terms of what the university, the sponsoring agency and at least one industry, the machine-tool industry, derived from it; to emphasize the important role played by the US government in fostering the research and promoting results, and to suggest that the project might serve as a model for current-day cooperative research efforts among the government, universities, and industry; and to set forth the proposition that the university, through its openness and skills in the educational domain, is an excellent environment for the conduct of engineering research, especially when there is no foreseeable need for the research results.

An attempt has been made to present the material in a way that will make it of interest to professional historians of technology, students of history, corporate administrators, managers of engineering research projects, government policy makers and others in government who have responsibility for initiating and overseeing government-sponsored research, as well as engineers who would enjoy taking nostalgic look at the early days of technologies that are now solidly entrenched in the marketplace. However, the book is not the stuff from which motion pictures are made. It contains no comedy, intrigue, drama nor confrontations, because there were none. The story is unfolded with no artificial coloring.

Characteristics of the technology innovation process, the scope of the research, and the environment in which it took place are presented in Chapter 1. Chapters 2, 4 and 5 cover the three phases of the effort—the experimental numerically controlled milling machine, the automatically programmed tool system, and the computer-aided design research. Chapter 3 is included for the benefit of those who wish to know technical details about the experimental milling machine. The roles played by the Air Force, Servomechanllisms Laboratory and industry are set forth in Chapter 6, and lessons that can be learned from the events that occurred over the twenty year span of the research are discussed in the final chapter. In an Afterword, the upward and downward turns of events that have transpired in the numerically controlled machine tool field since the US machine tool industry took over the technology are reviewed, and a case is made for the need for new research programs similar to the original NC project in order to restore the industry to a world leadership position.

J. Francis Reintjes

ACKNOWLEDGMENTS

Many persons inside and outside the Massachusetts Institute of Technology made important contributions to this book. Fortunately, several colleagues who were staff members of the numerical-control project, or on the periphery of it, were available for consultation as the narration unfolded. They provided personal recollections of project events and critiques of the manuscript. Included in this group were: Gordon S. Brown, William M. Pease, James O. McDonough, Alfred K. Susskind, Robert W. Mann, John E. Ward, Douglas T. Ross, and Jay W. Forrester.

William M. Webster and Max A. Guenther, Air Force civilian engineers who oversaw the Air Force's administration of the project, graciously read the manuscript for accuracy and provided valuable inputs. The comments of Edgar L.McFerren, formerly vice-president of the Giddings and Lewis Machine Tool Company, on an early draft were most helpful and were used in later refinements of the text. The Association for Integrated Manufacturing provided material on the origin of its Society.

Within the Institute, Paul E. Gray, Chairman of the MIT Corporation, took time from his very busy schedule to read the entire manuscript and make many thoughtful comments and suggestions. Professors Joel Moses and Richard B. Adler, Head and Associate Head, respectively, of our electrical engineering and computer science department, were most supportive of the entire writing project; their encouragement is greatly appreciated.

Photographs in the book were supplied through the courtesy of the MIT Museum staff, Cross and Trecker, the Giddings and Lewis Machine Tool Company, and the Aerospace Industries Association. Katherine W. B. Eyre and A. Lucy Allen rendered valuable assistance through their typing of various versions of the manuscript.

CONTENTS

NUMERICAL CONTROL

1
INTRODUCTION

This book is about issues and strategies involved in bringing new technology from the conceptual stage to its application to new products. Several revolutionary technologies that emerged from engineering research at the MIT Servomechanisms Laboratory from 1950 to 1970 are used to illustrate the process. Central to all of them was a rather unorthodox method of controlling machine tools automatically through use of digital-computer techniques. Eventually, the technology became known as *numerical control*, or simply NC. Although NC started out as real-time digital-control concept for machine tools, its underlying principles were gradually extended to other applications, so that now they are used widely for controlling many types of operations in the chemical, mechanical, electrical, and aeronautical industries.

The document goes beyond a simple chronological narration of events that transpired over the twenty or so years of the over-all research program. One reason for preparing the text was to record events from the viewpoint of one who participated in many of the important decisions made during the evolution of the technology, and to do this while most of the laboratory staff, and government and industry representatives who contributed to the effort were available for consultation. This kind of documentation was deemed important since it is not always possible for historical researchers to get first-hand accounts from those who made the history.

A second motive was to provide insights to a university-based research effort that has been judged exemplary by others, when evaluated in terms of what the university, the sponsoring agency (the US Air Force), and the machine-tool industry derived from it. There were beneficial fallouts for all participants. The Air Force achieved its goal of obtaining a new manufacturing method for its future aircraft, the machine-tool industry added a new revolutionary product line to its traditional lines, and MIT had an opportunity to pioneer a piece of forefront research in which its faculty and students played a prominent role. The research contributed significantly to the educational programs of MIT and other educational institutions.

Still another purpose was to emphasize the important role played by the US government in initiating and sponsoring the research, and the influential role it played in bringing together the university researchers and the industrial and user communities for a common understanding of the true merit of the technology. Indeed as the United States strives to preserve its prominent place in global markets through generation of new products, the numerical control research effort might well serve as a model for what can be accomplished through cooperative projects among the government, the university, and the private sector.

The numerical control project serves as an excellent example of what must be overcome in order to establish a revolutionary technology in the marketplace when there is no *a priori* demand for it. The book sets forth the proposition that in this kind of situation, university-based researchers, through their skills as educators and because of the open environment in which they work, can hasten the technology-transfer process.

As a preliminary to detailed descriptions of how numerical control evolved, it is appropriate to discuss the technical innovation process in general terms. This topic is the subject of the next section.

The Technology Innovation Process

The path from startup of an innovative technological research effort to the establishment of a new product in the marketplace is oftentimes long and unpredictable. This is especially true when the technologies are new and being used to synthesize a first-of-a-kind system. Further complications arise if the end result is a device or system of devices that performs a conventional task in a highly unconventional manner. Reorienting traditional thinking and establishing a strange operational mode as new standard practice for end users can then be a long expensive process with no assurance of success.

Engineering research, which is at the heart of the technological innovation process, is a commonly used term for investigative efforts that culminate in innovative methodologies for synthesizing systems, devices, or components. The term distinguishes the work from *basic* or *pure* research whose goal is the generation of new knowledge without regard for its practical, utilitarian value. Engineering research and applied research are synonymous but differ somewhat from *development*, which connotes a specific implementation among many others that might emerge from a piece of engineering research. Nevertheless, engineering (applied) research and development are closely related and are often intertwined, since many projects begin with a research phase and gradually blend into a development phase as the work progresses.

Many factors influence the course of engineering research and its outcome. Some rest on the degree of readiness of the technology itself and the creativeness of those who exploit it. Others are nontechnological in nature and include matters such as the soundness of engineering judgments rendered at critical points in the research process; the determination of management and administrators to see the program through in the face of setbacks and their willingness to supply the financial resources for doing so; unforeseeable happenings along the way and decisions made in light of those happenings; and sometimes simply good luck.

The expertise of the research personnel and their research style will have a major influence on the outcome of the effort. For an innovative, engineering-oriented project to succeed, a strong, dedicated, creative professional staff, capable of making sound engineering judgments—often on the basis of incomplete data—is an essential ingredient. Frequently, successful innovations have depended not so much on brute force use of new technology as they did on the imagination employed in its utilization. The staff must, of course, be at least aware of and preferably expert in all the technologies that might be involved.

In the realm of automation it has been amply demonstrated that automating a manual process simply by replacing manual operations with automatic ones on a one-for-one basis is often ineffective. Better automated systems can usually be contrived by ignoring traditional manual procedures and re-examining the process in terms of how automatic subsystems and components can be synthesized to yield the overall end result. Hence, success rests in part on the research staff's willingness to invest the time required to identify basic requirements and the fundamental issues that underlie them, and its ability to meet requirements through selective use of those conventional and unconventional technologies that offer highest promise of success. It is here that conflicts can arise between the researchers and their administrative supervisor and between the administrative supervisor and the research sponsor. This matter can be especially troublesome when the goal of the administrator is to get the job done as expeditiously as possible, whereas the research staff wants to examine basic issues thoroughly before plunging ahead. Similarly, difficulties can arise when the sponsor, whether an insider or outsider, underfunds the effort while adhering to expectations that cannot be achieved with the funds provided. Underlying this kind of conflict is the disparity in the degree of technical expertise that frequently exists between researcher and administrator. Failure of the latter to appreciate the magnitude and technical complexity of the problem can lead to impatience with slow progress and eventually to a decision to terminate the effort.

On the other hand, the researcher may also contribute misunderstandings through failure to recognize that engineering research is applications-oriented and therefore demands a quicker tempo than that found in the realm of basic research. Prolonging the work excessively in the hope that "an even better way" can be found is risky; it may eventually lead to an administrative decision to stop and "cut losses." The engineering researcher is continually faced with the difficult choice between proceeding with a plan that can be expected to yield satisfying results and searching still further for a better plan that may never materialize. Professional pride tells him to press on; practical considerations require him to temper his idealism.

Even though the outcome of engineering research and the time required to achieve success are unpredictable at startup time, providers of research funds should not be expected to endow the research indefinitely. Periodic accountability is a valid requirement. Oral communications on progress and near-term plans, as well as written reports and memoranda, are effective mechanisms for avoiding misunderstandings and letting sponsors know what kind of return they are getting on their investment.

Research style plays an important role in determining the final outcome of the project, even though there are no rigid rules for best procedures. Some researchers prefer to start immediately with the so-called broad, general problem in its class, knowing that if the general problem is solved, specific subsets of it will also have been solved. Others prefer to work inductively by tackling subsets of the big problem first and progressing gradually to the more complex final requirement. The expectation is that partial results can be extrapolated upward to the general case. Sometimes choice of approach is out of the hands of the researcher; it is set automatically by terms of the research agreement. For example, a sponsor may insist on seeing results on a crucial aspect of a problem before making a major financial commitment to a longer term effort. The risk in this approach is that the technology employed is a simplistic demonstration of a subsystem may not be the best technology for the more complex general case; it may be necessary to scrap the initial approach and start over in order to solve the general problem. Another possible risk is that a subsystem solution, when repeated several times as the solution is generalized may result in an unrealistically unwieldy system which may have been avoided had the big system been tackled at the outset.

On the other hand, researching the broad general problem first has its pitfalls. The research has an all-or-nothing flavor which may create an air of suspense over time and generate increasing doubts about a successful outcome as the work drags on and on. Researchers who adopt this style run the risk of overtaxing the patience of their sponsor and having their funds cut off before tangible results can be demonstrated. Nevertheless, the rewards for success in the general approach are great. General solutions can usually be expected to have widespread applicability to a host of specific problems.

Another aspect of style involves the division of the research group's time among system study and analysis, designing and building subsets of the system, experimentation, testing, and performance evaluation. Again, there are no rigid rules on how time should be divided among the tasks, but there are risks in extremes. Moving quickly into a building phase in absence of a thorough system study and analysis of requirements can lead to false starts and unsatisfying results. On the other hand, a prolonged study of the problem may never turn up practical issues that will become apparent only through "breadboarding" crucial circuits, mechanisms, or subsystems.* Interplay between analysis and experimentation often proves to be highly effective in that results obtained through experiment indicate directions the overall system study should take or avoid; thus, pitfalls that would not be uncovered until much later can be bypassed. In reality, the technological in-

*The term *breadboarding* signifies a temporary assemblage of components that are put together for experimental purposes only. The term dates back to the early days of radio when experimenters would mount their radio components on slabs of wood that resembled breadboards (or on real breadboards borrowed from the kitchen).

novative process is a multistage process with many feedback paths from the later to the earlier stages of the ensemble. As the synthesis process progresses, new information can be injected at various stages to overcome observed deficiencies and improve performance.

Engineering research in a university setting brings with it a whole new set of opportunities and constraints. It should be noted that opinion is divided among engineering-school faculty as to whether or not applications-oriented research belongs in the university environment. Some contend that the search for new knowledge for knowledge sake, unfettered by practical constraints, is the university's true role. Others would argue that applications-oriented research that permits faculty and students to work together on authentic, real-world problems offers the highest level of preparation for those who intend to enter the engineering profession.

Ability to conduct meaningful engineering research in a university setting depends to a great extent on the discovery of worthwhile problems that can be researched within the confines of the academic research laboratory and on finding problems whose time requirements for solution are consonant with the pace at which faculty members and their students can move the work along. Academic research will most likely proceed more slowly than it would in a nonacademic setting, principally because of the divided responsibilities of the participants between research and classroom teaching or study. The dual commitments of the participants also means that the research must generally be performed on campus, a constraint that may limit choices of research problems. For example, in the early 1950s a group of wool merchants from Australia came to MIT's Servomechanisms Laboratory hoping they could fund a project on improving the efficiency of wool processing from animal shearing through the baling of the raw wool. They were interested in as much automation as possible in order to reduce costs and preserve their position in the marketplace in the face of intense competition from synthetic fibers. It was clear their choice of Servomechanisms Laboratory was appropriate from the viewpoint of technical competence in system studies and automation, but quite inappropriate in terms of the Laboratory's ability to put personnel on-site in Australia, where the work had to be performed. Faculty and student participation on a project at that location was obviously not possible. Since the business of an educational institution is the education of students, and since research performed by students is recognized as a valuable form of education, it follows that a measure of worth of a university research project is the degree to which students are involved in the effort.

Nevertheless, there may be extraordinary reasons for undertaking an off-campus research effort. For example, an opportunity may arise which poses an unusual intellectual challenge, or offers potential for a significant contribution to engineering knowledge, or fulfills a major national need. Such a project may be justifiable in a university setting if some of the tasks can be performed oncampus by faculty and students. The remaining ones can then be handled by full-time research staff members assigned to the remote location.

Engineering research in the university environment brings with it additional opportunities to exercise the university's educational role. Being an open environment, free of proprietary constraints and industrial-type competitiveness, it is easy to apprise outside organizations of progress and to invite outside guests to participate in the activity on a temporary basis. This cooperation between university and industry can serve as a valuable mechanism for transferring research results expeditiously from laboratory to industry, as well as another way for the university to fulfill its educational role.

In industry the process of moving research results from laboratory to the market-place is rather straightforward. More often than not in US industry it is the work product that moves through the organization's research, engineering, production, and marketing departments rather than the personnel who perform the original work at the research stage. In other words, researchers do the research, design engineers design products based on the research, the marketing group does the marketing, and so forth. However, because the university's sole function is to educate, there is no mechanism within the university itself to translate its research results into useful products. Hence, when the time comes to move a process or a system outside, means must be sought for this purpose.

Several mechanisms are available for transferring new technology from the academic research laboratory to the industrial sector; each has its own advantages and disadvantages. Publication of the results as technical articles or reports is a common one, since it is consonant with the scholarly tradition of the university and its charter to disseminate information widely. However, a link in the technology transfer chain may easily break at this point if no one chooses to pick up reported results and build upon them. Another possibility consists in the academic laboratory joining hands with an industrial organization in order to refine the process, device or system to the point where the industrial group is able to take over and finish the job. This approach ensures continuity but may engender complications at the university laboratory because of the extra time involved. Because the strength of a university is measured in part by the magnitude and profoundness of its research programs, strong encouragement is given to faculty members to move out of mature projects and into new ones as quickly as possible. There is an ever-present risk to the researcher's long term future by staying with a project too long. A successful middle ground at the MIT Servomechanisms Laboratory has been to invite industry to join with us at the original research stage, with the expectation that our collaborators will acquire sufficient expertise to reduce the research results to practice as soon as the research is completed.

Overview of the Research Program

Three interrelated engineering research projects, all conducted at the MIT Servomechanisms Laboratory, have been chosen to illustrate the principles and issues set forth in the preceding section. They are: the development of a continuous-path numerically controlled machine-tool system; a high level, special purpose programming system for numerically controlled machine tools; and a set of innovative projects on computer-aided design for mechanical parts intended for machining by numerical control means. Together, these projects covered a 20-year time span from 1949 through 1969.

The initial phase of the research ended in 1953 and culminated in a laboratory demonstration of a digital electronic control system that simultaneously controlled the motions of a three-axis milling machine. The control technique became known as numerical control. The next phase might be appropriately called the transitional phase. It involved the transfer of numerical machine-tool-control technology from the laboratory to industry and the refinement of the numerical control concept through development of a special purpose computer programming system known as the automatically programmed tool (APT) system. From 1960 through 1969—the computer-aided design (CAD) phase—we investigated the role of digital computers in those aspects of the overall mechanical parts-design and parts-fabrication process that precede the use of digitally controlled metal-working tools.

The overall mechanical-parts manufacturing process is illustrated pictorially in Fig. 1.1. The top diagram in this figure shows the traditional method as it was practiced at the time the project began. The designer executed his design at a desk or worktable and interacted with a draftsman who translated the final design into a set of working drawings, or blueprints, so called because the reproduction process used to replicate black line drawings yields white lines on a blue background. The information stored on the blueprints was then translated into the required part by a machinist skilled in manual operation of machine tools, or alternatively, through use of templates that could automatically guide a tool along predetermined paths.

During the first phase of our project, the traditional design part of the process remained intact, but an automatic digital electronic controller—essentially a special purpose digital computer—was developed to replace the machine tool operator or the template guidance system. Hence, two new elements were introduced: the digital controller and a person known as a part-programmer, who translated the information stored on blueprints into numerical data that could be understood by the digital controller. Thus, the new sequence of operations in the overall manufacturing process followed that illustrated in the second row of sketches shown in Fig. 1.1, and the new machining process was called a numerically controlled process.

In the beginning, the digital controller was hand programmed with the aid of a mechanical calculator, a procedure which proved to be slow and tedious—especially when an odd-shaped three-dimensional part had to be sculptured from a block of metal. In an early attempt to streamline the programming task, a general purpose computer was utilized as a more powerful aid in making the required mathematical calculations. A set of computer programs, called library subroutines, was developed and organized so that they could be selectively called upon to make those calculations which were likely to be common to every job.

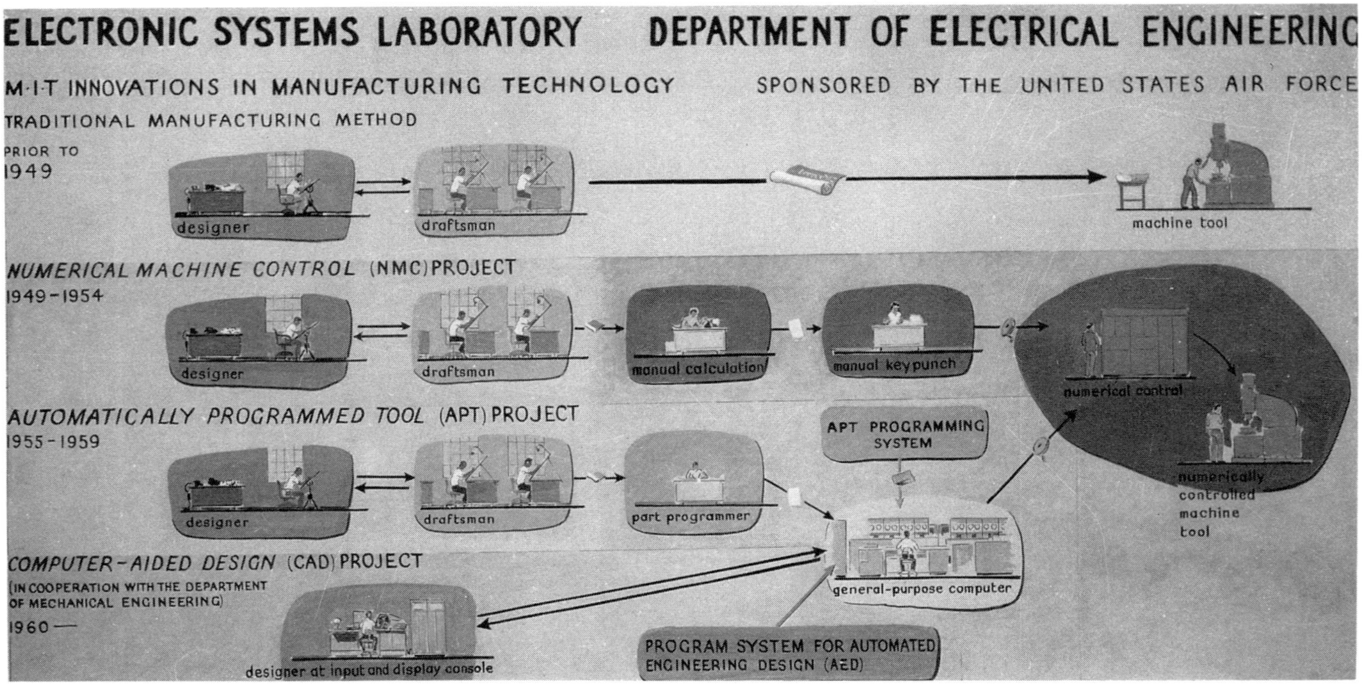

FIG. 1.1. Pictorial representations of the various stages of the MIT numerical control project.

Although the computer library subroutine approach was an important forward step (for a short time one aircraft company actually made limited use of it), the concept fell short of what was ultimately needed to put programming time and cost on a firm economic footing. It was during the second phase of our project that we tackled this problem in a major way and were able to come up with a solution that turned over most of the programming drudgery to a general purpose computer. A set of special purpose computer programs was developed that could temporarily convert a general purpose computer into a powerful special purpose computer to perform automatically the computations that formerly had been done by hand. As a result, the role of the part-programmer was greatly simplified. Machine tool instructions could now be stated in simple, English-like expressions, and the computer could process these instructions and generate a series of numerical commands to guide the machine tool in accordance with the original instructions. The system of computer programs that emerged from this phase of the project was called the automatically programmed tool system, commonly abbreviated APT. In the presence of APT, the manufacturing process operated along lines shown in the third row of the diagrams in Fig. 1.1. Observe, however, that the designer's and draftsman's roles still remained as they were at the start of the project.

Servo Lab's CAD research evolved as the result of the experience gained during the two preceding phases of the project. Gradually it had become evident that ability to control machines by numerical methods had implications far beyond the machine itself. The overall design and manufacturing process could be thought of as an *information processing* sequence, which in many instances could be carried out effectively through use of digital (numerical) methods. That being so, the whole chain of events that occurs between the initial concept of a mechanical part and its physical realization could be carried out with the aid of information processing systems. In the world of computer-aided design, the designer would be able to make use of the digital computer as a design tool to execute his design, and instructions to the metal-cutting or metal-forming machine for implementing the design could also be conveyed in numerical form.

However, if the design process were to proceed in this new direction, powerful computer-programming systems capable of handling the many complex issues associated with creative design would be needed. There would also be required new kinds of electronic display terminals that would serve as interfaces between the computer and designer and allow the designer to work back and forth with the computer as the design progressed. These two broad tasks constituted the third phase of the research program. They culminated in a family of special computer programs known as the automated engineering design (AED) programming system and in powerful electronic graphical display terminals that enabled the designer to draw and manipulate physical shapes and devices on cathode-ray tube screens. In terms of Fig. 1.1, the computer aided design and manufacturing process would be carried out as shown on the fourth line. The designer would interact with a general purpose computer as the design developed, and the general purpose computer would supply design information to the specially programmed APT machine, which would then generate instructions for the machine tool.

The Servomechanisms Laboratory

The environment in which research is conducted plays an important part in determining the success of the work. The goals of the laboratory should be consistent with the character and goals of the research in order to avoid conflicts between researchers and administrators. Doing applied research in an environment dedicated to basic research, for example, can prove frustrating to both parties. The laboratory administrators should have an understanding of the magnitude of the problem at hand, including its technical ramifications, and an appreciation of the time that might be required to fulfill terms of the research agreement. It is highly desirable, also, that the project leaders be given a free hand to exploit their own ideas and to carry out the work in accordance with their own style.

Fortunately, the environment in which numerical control took form was ideally suited for this undertaking. The excellent match that existed between the goals of the project and those of the Laboratory contributed significantly to the success of the work. The research was carried out in the Servomechanisms Laboratory —commonly called Servo Lab—which was one of several research laboratories in which MIT's Electrical Engineering faculty members conducted their research. Founded in 1940 by Dr Gordon S. Brown, then Assistant Professor of Electrical Engineering, the Laboratory initially undertook research and development in the application of servomechanisms to military problems, particularly in the realm of gunfire and guided missile control. However, it soon expanded into areas such as flight simulators, digital computers, and controls for an experimental nuclear reactor. Two important projects of the middle and late 1940s, for example, were the Whirlwind computer project (1–1)* and the control system for the Brookhaven National Laboratory reactor for nuclear research. Housed in a 20,000 square-foot, single-story building at the rear of the campus, the Laboratory was often referred to as "one of those research labs on the back lot." It was also known as Building 32, in accordance with MIT's practice of denoting its buildings by numbers.

Organizationally, the Laboratory operated within the Electrical Engineering Department, with its director reporting to the Departmental Head. Servo Lab had its own machine shop and its own purchasing, accounting and drafting departments, but in accordance with the MIT policy of administering research contracts and grants, it reported to the Institute's Division of Industrial Cooperation (DIC). First established prior to World War II, with Nathaniel McL. Sage as its head, DIC underwent two name changes over the years, to the Division of Sponsored Research (DSR) and later to

*Numbers in parenthesis refer to items in the list of references at the end of the book.

Gordon S. Brown

the Office of Sponsored Projects (OSP), but its role has remained the same: to provide linkage between the Institute's researchers and its research sponsors, and to provide administrative assistance on matters pertaining to research grants and contracts.

Brown's research in servo applications nicely complemented his classroom teaching activities in feedback control, a new technological area that had emerged in the 1930s.* The tone of the Laboratory's activities was set by Brown's enthusiasm to take on authentic, challenging engineering tasks and to get the job done in an atmosphere that brought together faculty and students as a research team. Throughout the fifty years of the Laboratory's existence, students have played a strong role in its research, and for many years its theme continued to be "engineering research backed up by fundamental studies"—when such studies were required to provide the sound underpinnings for the engineering-oriented projects (1–2).

The Laboratory has always operated on a project basis rather than through a single, blanket research grant encompassing the entire spectrum of its activities. Usually, each project consists of a faculty member or a senior engineer serving as group leader, assisted by graduate and undergraduate students, and sometimes one or more junior staff engineers selected from the Laboratory's graduate students upon completion of their graduate studies.

*In the concept of feedback control, a signal from an output member of a system of components is *fed back* to an input member for the purpose of making a comparison between input and output signals. Any error that exists is amplified and used to bring input and output into close correspondence, thereby reducing the error in desired performance to a minimum. A servomechanism, or servo, is a system of electrical and mechanical or electrical, hydraulic and mechanical components that includes feedback.

Junior staff seldom have remained at the Laboratory for more than a few years, principally because their work in the Laboratory often opens up highly attractive career opportunities for them in industry.

In addition to those already mentioned, a small sample of projects that have phased in and out of the Laboratory over the years includes such diverse topics as guidance and control systems for underwater vehicles; feedback control for chemical processes; an automatic pressure control system for an iron lung used in research on and treatment of poliomyelitis; solid-state subsystems for aircraft and aerospace radars; a computerized traffic control system for the Eisenhower tunnel in Colorado; investigations of situation displays for cockpits of commercial airliners; analyses of command, communications and control systems; computer applications in the newspaper business; large-scale information storage and retrieval systems; and investigations of solid-state scanners for binary encoding of printed materials. In addition, throughout its fifty years of existence the Laboratory has enjoyed an international reputation as a leader in the advancement of theory relating to the dynamic performance of linear and nonlinear systems. It was this academic engineering research environment that nurtured the numerical control projects as they moved through the three phases already cited.

With such a broad array of research topics, the name Servomechanisms Laboratory was hardly descriptive of its activities, and could even be construed as a misnomer. Indeed this was recognized in 1959 when the name was changed to Electronic Systems Laboratory, and in 1978 when it was changed to the Laboratory for Information and Decision Systems.

Milling and Milling Machines

Milling is a method of removing metal by means of a rotating cutter. Machines that perform milling do so through their cutter action and simultaneous motions along one or more of the three orthogonal axes of the machine. The blank stock is held in place on the machine's worktable by means of a jig or fixture, and the table is power-driven along two of the three principal axes. The cutter spindle, which is horizontal in some designs and vertical in others, is driven along the third axis.

Milling is a continuous process, meaning that during its operation the workpiece is contoured continuously to the desired degree of accuracy. This characteristic is in contrast to discrete-positioning machines—drilling machines, for example—where accuracy is required only for the location of the holes. The path the cutter takes while moving from hole to hole is unimportant since cutter and workpiece are disengaged during that interval.

Milling machines are manufactured in a variety of sizes and with features that make them suitable for use in general manufacturing or for special types of jobs. A general purpose machine is illustrated in Fig. 2.2, and five-axis skin milling machines, specially designed with two cutting heads and used to hog out aircraft wing panels from aluminum slabs, are shown in Fig. 4.2 and Fig. 6.5.

Tracer-controlled milling machines were developed in an effort to improve accuracy, reproducibility and reliability, particularly in the sculpturing of complex shapes. These machines were in common use at the time our numerical control development effort was initiated. In tracer control, a pattern (template) that defines the desired contour in one or more dimensions is first made; it is then used with a mechanical stylus, or follower, as the means for automatically controlling machine motions in order to reproduce the pattern.

2

THE EARLY YEARS

How It Began

A doctoral thesis submitted in the spring of 1949 to MIT's Department of Electrical Engineering by William K. Linvill was instrumental in part in bringing the numerical control project to the Servomechanisms Laboratory. Linvill's thesis, entitled "Analysis and Design of Sampled Data Systems" (2–1), grew out of his work with C. Robert Wieser, a staff member of the Whirlwind project, in their planning for servomechanisms equipment capable of receiving pulse-type data from the Whirlwind computer, which was under development in the Laboratory. Linvill's thesis supervisor was Professor Ernst A. Guillemin of the Electrical Engineering Department. At the time, the theory and design of linear feedback control systems operating under the influence of signals that varied continuously in time were well understood; lacking was a similar understanding of such systems when input signals were discontinuous, that is, when they were in the form of pulses like those derived from a digital computer.

Professor Gordon S. Brown was well aware of the implications of Linvill's thesis research, particularly its relationship to digital computers, since it was in his Laboratory that the Whirlwind project got started, and he had served as one of Linvill's faculty thesis readers. Brown was therefore prepared for opportunities to merge in a single application the technologies of digital feedback control and digital computation.

It was by pure chance, in the spring of 1949, that Brown received a telephone call from John T. Parsons, who was vice president of the Parsons Corporation, Aircraft Division, in Traverse City, Michigan. Parsons was visiting the Wyman–Gordon Company in Worcester, Massachusetts, and Brown later concluded that the call was prompted by Clinton C. Lawry, Jr, one of Brown's former students. Parsons told Brown that he was looking for a power drive of a couple of horsepower for a machine tool application in which the input to the drive would be in the form of intermittent data. Brown, equating "power drive" to "power servomechanism," sensed this might be an opportunity to apply concepts of digital feedback control, so he agreed to explore the matter with Parsons.

For many years prior to 1949, it was common practice to contour a two-dimensional irregular surface by cutting closely spaced holes along a path that followed the desired contour and then finishing the contoured surface to the specified smoothness either by machine or manual means. Since the cutting edge of the tool was tangent to the desired contour, allowance for tool radius had to be included in the calculation of workpiece position for each cut. Cams and templates were often made by this process. Typically, the machining steps are these: the class of machine tool is chosen—a vertical spindle milling machine, or a jig borer, for example. The number of points along the contour is selected, and a reference point for specifying the coordinates of each point on the workpiece is chosen. The settings of the two cranks controlling the orthogonal movements of the table are then calculated to give the required coordinates of each point, and the workpiece is aligned with the reference point. The workpiece is first cleared free of the cutting tool, and the table is positioned to the coordinates of the first point. The tool and the workpiece are then engaged, the cut is made, the workpiece and cutter are disengaged, settings for the next point are cranked in, and the process is repeated. Note that the process utilizes a discrete-position technique in a situation that ideally calls for continuous cutting along an irregular path. Hence, there is a penalty in cutting time—no metal is removed during reset intervals—and extra work is required in order to bring the rough surface to the desired degree of surface finish.

During the 1940s, John Parsons had experimented with this discrete-position machining method to produce a template for an airfoil shape. Using two hundred discrete points to describe the shape, he was able to produce a template to the standard accuracy of ±0.007 inch, with no subsequent filing (2–2). Prior to the time he contacted MIT, Parsons was aware of the possibility of storing information on punched cards and later converting the stored data to a series of electrical pulses as the cards were read out. His thought was to store machining information on cards and use the pulse-type signals generated as the cards were read to drive power devices that would control the motions of a machine tool. In terms of an application, Parsons recognized the possibility of expanding automatic point-by-point position control to three dimensions for the purpose of making aircraft tapered wing panels. He concluded that this type of job would "require the positioning of five elements of a machine tool including longitudinal table travel, and front-to-back and vertical cutter travel as well as both elements of a universal table" (2–3). At the time Parsons was concentrating on the machining of tapered wing sections because he believed that existing machines would unlikely satisfy Air Force needs because they would be unable to produce these sections efficiently.

Parsons was encouraged to pursue his ideas by representatives of the Air Force at Wright–Patterson Air Force Base (WPAFB) in Ohio, and they had visited his shop for a demonstration—which proved to be a tedious one, since table positioning was accomplished by manually moving lead screws until the calculated location of each point was attained. Parsons, however, envisioned that the process could be automated so that coordinate information could be stored on punched computer cards and the lead screws driven automatically from data read from the cards. He dubbed his system the Cardamatic milling machine. At first there was doubt that a machine tool could be made to arrive at a specified coordinate position repeatedly within an accuracy of a few thousandths of an inch, but through his contacts at the Snyder Tool and Engineering Company in Detroit, Michigan, Parsons was able to show, in another demonstration, that such repeatability *was* possible. Parsons followed up the demonstrations with a formal proposal to the Air Force; it culminated in his contract AF33 (038)–6878.

At this point in time Parsons was evidently thinking in terms of an automatic machine that would do cutting continuously along straight lines in order to generate flat surfaces (or those that curved in one dimension only) and that would do plunge cutting to generate compound curved surfaces. In Parsons' words: "It was proposed that this equipment would be capable of straight line generation of surfaces only, since external wing surfaces could be produced by this method, and that curved surfaces such as the inside of integrally stiffened sections would have to be machined by a series of plunge cuts indexing the left to right and front to back motions of a machine tool to the required dimension and then making a vertical cut to the prescribed depth" (2–3).

The ultimate goal of the Air Force's contract with the Parsons Corporation was to "furnish labor, material, and services necessary to design, engineer, build, test, and install on a 6'×6'×10' standard planer or planer mill, the equipment necessary to perform automatic contour cutting, external or internal, upon surface contours similar to those found in a wing section . . ." The goal was to be reached through performance of four sequential tasks: (1) a design study of a servomechanism actuated by punched cards or tape which would operate a lead screw; (2) construction and test of a working model of a servomechanism to operate a work table approximately 18"×40", mounted on ways so that an adequate determination of performance and accuracy could be made under load conditions; (3) an application study and detailed design of a 3-axis machine with a work capacity of 18"×24"×48" and capable of machining both aluminum and forging-die steels; and (4) construction and test of the machine designed in (3). The contract amount was $200,000, allocated as follows: Task one, $50,000; task two, $40,000; task three, $20,000; and task four, $90,000. It was stipulated that startup work on each task, commencing with task two, would depend on satisfactory completion of the preceding task. The time allowed for completion of all four tasks was twenty-one months.

The Parsons–Air Force contract clearly was for proof-of-feasibility studies reinforced with some experimentation, rather than for a prototype of a production machine. What the Air Force sought was information on how well such a system could perform, presumably in terms of cutting speeds and improved accuracies attainable under load conditions. Evidently the Air Force took a conservative view toward the project: included in the contract was the right to "cut its losses" through a stipulation that startup of each task beyond the first would depend on a satisfactory completion of the preceding

William M. Pease

task. Just what constituted "satisfactory completion," however, was left open. The contract wording is also interesting in that a study of a pulsed-servomechanism to do the job was given first priority. This fact, together with the right-to-terminate clause, can be construed as an indication that the Air Force wanted to know as much as possible about the power drive system before it made a further commitment of funds. This is understandable, of course, since relatively little was known at the time about the design and performance characteristics of pulsed-servos.

An Early Meeting at Servo Lab

Shortly after Parsons' call to Gordon Brown, a meeting was held at Servo Lab (in late June of 1949) to explore a possible contractual relationship with the Parsons Corporation. Participating in the meeting were Parsons, Professor Brown, and William M. Pease, a Servo Lab staff member. Parsons described the machine tool he had in mind and outlined the general terms of the contract he had received from the Air-Force to pursue his idea. He told Brown and Pease that he wanted to explore the possibility of Servo Lab's taking on the job of designing and building the servomechanism described in his contract; input data for the servo would be derived from somebody else's punched cards or tape. As the meeting continued, Parsons unfolded his overall plan for carrying out the contract. Since the Parsons Corporation had neither facilities nor professional staff with experience in the forefront areas of digital electronics and pulsed feedback control systems, it would have to depend on outside organizations to develop the tool and controls along the lines Parsons had in mind. At the time of Parsons' visit to Servo

James O. McDonough

Lab, IBM (Endicott, New York) had already been asked to look into issues related to the use of information stored on punched cards as means for actuating power drives. Parsons also intended to ask Snyder Tool and Engineering to consider the possibility of supplying a demonstration machine. Thus, the Parsons plan was to utilize three outsiders—IBM, Servo Lab and Snyder Tool, each acting independently—to develop the information handling process, power drive, and machine tool components of the system. The weaknesses or this plan, of course, were that no individual or organization had been hired to do a system study and an analysis of the overall Cardamatic concept, and to provide technical leadership, coordination and direction to the subcontractors.

Nevertheless, the Servo Lab staff was challenged by the possibility of developing a digitally driven servo for a specific application, so Brown agreed to take steps that would lead to a contract between MIT and the Parsons Corporation. In response to John Parsons' inquiry about the cost of a study of the pulsed servomechanism problem (outlined in task one of his Air Force contract), Brown recalled: "I told him about $25,000 was a round number for a power drive of the kind envisioned." The $25,000 figure corresponded to approximately one man-year of project effort at that time. The meeting closed with the understanding that Servo Lab would begin preparation of a document that would serve as a basis for a formal contract with the Parsons Corporation.

MIT's Agreement with the Parsons Corporation

Following the first meeting with Parsons, Brown put Pease in charge of the impending project, and James O. McDonough was assigned to assist him. Pease had received the SB and SM degrees in electrical engineering at MIT six years earlier, in 1943. He had performed his master's thesis research, on the design of a 400-cycle servomechanism, under the guidance of Professor Albert C. Hall of the Electrical Engineering Department, also a staff member of Servo Lab. Upon completion of his graduate studies, Pease became a research staff member of the Lab, where he served as project engineer of the Brookhaven National Laboratory reactor project, another Servo Lab effort. On July 1, 1949, he joined the faculty as Assistant Professor of Electrical Engineering.

McDonough was also an MIT alumnus, having received his MS degree in electrical engineering in 1947. Professors Brown and Pease were co-supervisors of his graduate thesis: "The Dynamic Characteristics of a DC Motor with Quadrature Axis Compensation." Upon receipt of his graduate degree, McDonough joined Servo Lab as a full-time staff engineer assigned to the Brookhaven reactor project for which hydraulic servos were being developed.

Given their backgrounds and experience in servomechanisms, Pease and McDonough were logical choices to pursue the Parsons Corporation's servo requirements for the Cardamatic machine. Besides, since the Brookhaven work was coming to a close, they were available for reassignment. Pease and McDonough's approach was to learn as much as possible about the machine tool control problem as quickly as possible. To assist them in this endeavor, they drew heavily on the skills of other Servo Lab personnel familiar with servomechanism techniques and on the knowledge of their colleagues working on the Whirlwind digital computer project. At first the only information available for them to work with was what could be deduced from the broad specifications conveyed verbally by Parsons during his first visit. However, upon receipt of its Notification of Award from the Air Force, the Parsons Corporation began putting together its own set of preliminary specifications for what it called the Cardamatic Milling Machines I and II. The Parsons Corporation specifications called for two designs: an experimental working model and a final machine. The latter would have a maximum longitudinal table feedrate of 180 inches per minute for straight-line cuts—three times faster than the experimental model—and minimum and average rates of 0.5 inch and 6 inches, respectively. Table dimensions would be 28"×72", in contrast to 18"×40" for the experimental model (the version described in the Air Force document). These specifications were given orally to Servo Lab personnel at a meeting held in Professor Brown's office on July 11, 1949. (A copy of the four tasks specified in the Parsons–Air Force contract was transmitted to Servo Lab staff at a later time.) Representing the Parsons Corporation at the July meeting was Robert Marsh, who had been hired to coordinate the work of the subcontractors. In addition to Gordon Brown, Servo Lab was represented by William Pease and John O. Silvey; they were joined by Robert R. Everett and C. Robert Wieser of the Whirlwind project.

It is clear from a July 12 memorandum (sent the day after the meeting) from Everett to Jay W. Forrester, leader of the Whirlwind project, that the

Jay W. Forrester

Servo Lab group was now trying to reduce some rather vague generalities to specifics, and that the group was beginning to identify the technical issues that would have to be resolved in order to meet the detailed specifications set forth by Parsons. From Everett's summary of the meeting, a serious design problem was seen as being posed by the need to convert decimal data from punched cards to digital electrical signals for running the servos that had to meet a positional accuracy of 0.003 inch in 6 feet. Also, the importance of a high degree of reliability emerged from Everett's simple statement: "The cards can be checked beforehand, but a control failure could ruin a wing or die worth $100,000." Although IBM cards were specified in Parsons' preliminary specifications, Robert Marsh of Parsons Corporation raised the possibility of storing information on magnetic tape as a buffer to a standard paper tape reader. Evidently Marsh was concerned about meeting the required cutting speeds, because of the slow rate at which punched cards could be read. Compounding all these issues was the observation made by Everett that: "The [Parsons] preliminary specifications are more nebulous than shown [on the attached sheet]; Parsons would like higher feed rates and will eventually go to larger machines and add rotatable and horizontal spindle heads as well as the vertical head listed." (2–4) At this point, John Parsons evidently was thinking beyond his Air Force commitments to a time when the utility of his machine would be judged in the marketplace by its cutting speed and versatility.

In light of the "moving target" nature of the specifications Servo Lab was expected to meet, the staff concluded that its first task was to undertake a thorough *system* study of the whole digitally controlled machine tool concept—something that had not as yet been done. It was not clear which of the several sets of specifications that were floating around had a reasonable chance of being satisfied by current technology. Consequently, MIT urged that such a study be made a part of any agreement with the Parsons Corporation.

One outcome of the July 11 meeting was the decision that Pease and Marsh would visit IBM the following week to see how that company was progressing on the card reader and digital signal processing aspects of the Cardamatic concept. A meeting was held at IBM's Endicott, New York facility on July 19, 1949; what transpired there is recorded in a memorandum from Pease to Gordon Brown, written the next day:

1. Mr Daygar [of IBM] outlined the work that IBM had done based on their early conceptions of the machine tool that Parsons Corporation was planning to build. The machine, more resembling a jig borer than a milling machine, was to execute individual operations in response to information furnished from a standard IBM calculating machine (Type 604). IBM was to furnish a card reader that would place in ordinary revolution counters the information furnished by the cards. Geared to each of five revolution counters were to be three synchro pairs for transmitting the information to the machine tool. Five table and spindle drives were to be used to position properly the work and cutting tool in the machine. The card reader was described briefly as a device to read a number from the card and place the number in storage, then compare the number in storage with the revolution counter and correct the setting of the revolution counter until the comparison showed no difference. As a check, the revolution counter was then to be compared with the card itself. A relaying system was planned to hold machine-tool operation at a standstill until card reading was complete.
2. Mr Marsh described the new specifications as set forth in the agreement between Parsons Corporation and the Air Force. Under this agreement the machine tool is to be similar in principle to a conventional milling machine with a horizontal table and a vertical spindle. The number of positioning drives is now three instead of five. Information is to be fed to the machine continuously. To obtain essentially continuous data, the computer will have to compute approximately 10 positions per second. Mr Daygar was asked if IBM Co. equipment could read cards at this rate. Mr Daygar estimated that the proposed device could read at approximately 2 per second and that as an ultimate limit the cards could be read safely at approximately 3 per second. Mr Marsh suggested that the cards be first read onto tape and then the tape run at high speed to achieve the rate of 10 per second. Mr Daygar agreed that the tape would be safer than the cards and suggested that IBM standard punched paper tape might serve the purpose. Mr Marsh suggested the possibility of using magnetic tape. Mr Daygar agreed that this was a good proposal, but so far as IBM was concerned, it would require a considerable development. At this point Mr Daygar emphasized the advantages of using standard IBM equipment where possible.
3. The general attitude of IBM as expressed by Mr Daygar seems to be willingness to furnish standard or modified equipment for this job, but reluctance to enter into any appreciable development work.
4. After the conference Mr Marsh informed me that the agreement being worked out between M.I.T. and the Parsons Corporation would be mailed

> by Mr Parsons within the next few days. So far as Mr Marsh is concerned, this agreement carries with it the understanding that we will continue investigating the complete system in order to arrive at the best possible solution for all concerned. (2–5)

Several noteworthy observations can be made about Pease's meeting at IBM. One is that John Parsons apparently had gotten IBM interested in his idea before he wrote up the specifications he gave to Servo Lab. (IBM had been updated on those specifications at the July 19 meeting.) Another is that the meeting brought out the fact that the punched cards Parsons had believed adequate for storing machine information were inadequate for the task because card readers then available from IBM could not accommodate the cutting speeds Parsons had in mind. Moreover, the ever-present possibility that a stack of cards might be dropped on the shop floor, or elsewhere, did nothing to promote confidence. When the possibility of substituting punched paper tape or magnetic tape as a storage medium had emerged, it was pointed out that either alternative would entail considerable development work, which, as Pease concluded, IBM was not about to undertake.* On a different point, an important fallout from the IBM meeting was the assurance given to Bill Pease by Robert Marsh that the proposed MIT—Parsons agreement, about to be signed by Parsons, carried in it the understanding that MIT would conduct a complete system investigation in order to arrive at the best possible system configuration.

The IBM meeting was an early turning point in the history of the numerical control project. Although the Cardamatic concept as Parsons had conceived it was not dead at this juncture, major adjustments would have to be made if punched cards, the kernel of the concept, were to be retained as the information storage medium. As suggested at the July meeting in Endicott, New York, an alternative would be to read the cards into another storage medium, such as paper or magnetic tape, so that the required readout rates could be achieved. The disadvantage of this approach was that it would necessitate additional development work beyond that specified in the Parsons—Air Force contract; furthermore, IBM, the logical developer, did not want to undertake the work. Another possibility was to reduce the feedrate requirement to make it compatible with the card reader speed. The risk in this approach was that a digitally controlled machine might be proven technologically feasible but too slow to be acceptable to machine tool users. John Parsons was looking for faster rather than slower rates in order to make his machine attractive in the marketplace. Thus MIT began to search for other system configurations that could meet performance specifications and be implemented with currently available technology.

The agreement between the Parsons Corporation and MIT was signed on

---

*Even at a reading rate of 10 cards per second, cutting speed would be low. With the 0.003-inch positional resolution specified by Parsons, cutting speed would be 0.03 inch per second, or 1.8 inches per minute. The resolution that was attained with Servo Lab's experimental milling machine (at a later time) was 0.0005 inch and its maximum cutting speed was 15 inches per minute.

July 29, 1949. It was a simple, two-page document that called for a one-year project, starting July 1, 1949, at a rate of effort "to be established in a mutually agreed-upon budget which the Institute will not exceed without specific written authorization or modification of the budget to take into account new conditions." (The budget amount finally agreed upon was $30,000, but as the work progressed, expenditures increased to a total of $80,000.) The work was carried out in two phases, as follows:

Phase 1. An engineering study directed toward the development cf a system applicable to machine tools for controlling the position of shafts in accordance with the output of a computing machine.

Phase 2. The supply of an apparatus to operate a demonstration machine tool to hold a performance accuracy so that the surface finish has no roughness in excess of .001 inch, and is correctly located within .0004 inch when operating at a commercial table rate. (2–6)

The word "research" appears repeatedly throughout the agreement, and the open-ended character of the phase one and phase two statements gave MIT the directive it needed in order to explore alternative design approaches to digital machine tool control. (The complete Memorandum of Agreement between Parsons and MIT is shown in Appendix A.) The Agreement is noteworthy for several reasons. The salary and wage rates included an overhead cost of 100 percent,* which "enables our commercial work to make a contribution to the Institute to help it maintain its facilities and research position." It is significant also that the Parsons agreement with the Institute was not tied to any Air Force contract: it was solely between the two parties and the two phases of research were not in one-to-one correspondence with the tasks outlined in the Parsons—Air Force contract.

The First Year

The first few months of the project, July through October 1949, were startup months. As project engineer, Pease had the job of identifying the tasks that should be undertaken first and of putting together a team to work on them. McDonough's first assignment was an investigation of the mathematics involved in corrections to cutter tool center data when metal removal occurs at the edge of the cutting tool, and a study of circuitry that might be used to implement mathematical requirements. McDonough also began to study sensing techniques that might be included in the feedback loops of the servo drives. His goal was to understand the sensitivity and accuracy requirements for sensors that would measure the position of the machine tool table.

By the end of August 1949, the number of personnel on the project had doubled through the addition of Dr Ralph Kochenburger and Rudolph J. Cypser to the roster. Kochenburger, who had completed his doctoral dissertation on analysis and synthesis of contactor servomechanisms the preceding

*Salaries in those days were much lower than current rates: our staff engineers were paid $2.25 per hour, or about $4,500 per annum, an amount that was typical at that time!

June, was given the task of investigating pulsed-servo requirements and cutting-tool offset paths, while Cypser undertook a study of analog-to-digital conversion techniques—specifically, the conversion of shaft position to binary codes. During the fall of that year, J. Herbert Brown joined the project staff, but Kochenburger resigned at the turn of the new year and joined the the faculty of the University of Connecticut. Cypser's stay on the project was short, too—he switched to another project within the Lab in late Spring, 1950.

William Pease had begun the system study called for in phase one of the Parsons–MIT Agreement, and was able to devote full time to this effort during July and August 1949. Commencing in September, however, Pease had to divide his time between the project and his teaching and other academic commitments. Consequently, work proceeded rather slowly at first: one-quarter of the way through the contractual period only about 11 percent of the total contract amount had been expended. Nevertheless, the staff was well prepared for a major three-day meeting that took place at Servo Lab from October 24 through 26, 1949. Participating in this gathering were John Parsons, Frank Stulen and Robert Marsh of the Parsons Corporation, Leo Gayda of Snyder Tool and Engineering, and the Servo Lab staff. They were joined, on the second day, by an Air Force representative from Wright–Patterson AFB, C.A. Burdg. This meeting was to prove important for several reasons. It was the first time all the principals had gathered at the place where the research was being conducted and there were several issues that had to be settled: namely, how to proceed in view of the failure of the IBM card reader to meet the feedrate requirement and IBM's desire to pull out of the joint effort should development work be required. Also, since the tasks that the Parsons Corporation had worked out with MIT did not jibe with those specified in the Parsons–Air Force contract, John Parsons wished to bring these two documents into conformity.

The first meeting day was spent on a review of technical progress to date, a review of contractual requirements, a visit to the Whirlwind project to see digital data storage and paper tape reading equipment, and planning for the next day's meeting at which the Air Force would also be represented by C. A. Burdg. On the matter of contract revision, Pease indicated that he would like to see items one and two in the Parsons–Air Force contract merged in order to bring it into conformity with the way Servo Lab would like to proceed with its system study. He pointed out that "Servo Lab generally produces a final system design by performing basic studies, making partial designs, changing the design and hardware as needed, and ultimately converging upon a final design, all of whose features have been tested during the design process." This style would necessitate merging items one and two in the prime contract, so it was agreed that Parsons would formally request this change from the Air Force, which he did. Parsons also informed Burdg of other contract changes that should be made in light of events that had transpired since the award of the contract. Specifically, he described the advantage to be gained in cutting speed by using tape rather than punched cards as the information storage medium. Parsons urged that $75,000 of his

Air Force contract be earmarked for the development of a special tape reader and that MIT be given the job of developing it. Related to this point, Parsons expressed a desire to eliminate a contract item stating that IBM would develop a tape or card reader at no expense to the government. It was agreed that he would work with WPAFB personnel on the recommended contractual changes. As a result of these negotiations, Supplement No.1 to the Parsons–Air Force contract was written; it included an additional $75,000 for development of the tape reader.

During the two-day meeting, an interesting set of views emerged on how the project ought to proceed (2–7). There was general agreement that the system study should be conducted along lines advocated by Pease, as described above, but there was much discussion about the nature of the system that should be investigated. Both Parsons and Pease agreed that the long-term goal was to develop a three-axis, digitally controllable machine tool, as described in tasks three and four of the Parsons–Air Force contract, and Pease wanted to move immediately toward that goal. John Parsons, however, thought that a single-axis demonstration, using a machine tool table specially designed by Snyder Tool and Engineering, should be the basis for the study because it would satisfy his desire to demonstrate that a servo drive could move a machine axis under cutting loads and *stop* at a desired location—a one-axis machine would demonstrate what Parsons most wanted to demonstrate. He continued to cling to his ambition to build a specially designed machine tool that would sculpture one product only—the wing sections for supersonic aircraft. Although Pease did not rule out this narrow approach, being mindful that task two in the Parsons–Air Force contract called for a single-axis demonstration using a small worktable, he wished to make no firm commitments until the Lab had moved further along in its system study.*

Pease preferred to start with the assumption that digital storage, digital data processing, and pulsed servos would be available for controlling machine tool motions along three axes, and then study how these technologies could be most effectively employed to meet performance specifications of general purpose tools that could machine a variety of parts. In other words, Pease did not wish to immediately become locked into an approach that would meet short-term goals but which might not be extrapolatable to machine systems that would satisfy the Air Force's long-range goals. It was finally agreed that, to the extent that MIT's agreement with Parsons Corporation could be completed within the allocated budget, Pease would move along the line he advocated, and that a specially designed Snyder machine tool would be considered, along with others, for demonstration purposes. As will later be seen, a conventional commercial three-axis milling machine, a Cincinnati Milling Hydro-Tel, was finally chosen, and furnished to the project by the Air Force from its stockpile of machine tools.

*Pease commented many years later that, from the beginning, Servo Lab envisioned broader applications for digital control beyond one which would just sculpture aircraft wing sections through use of a specially designed machine tool.

Cutting speed was brought up again for discussion at the meeting, this time by the Air Force's Burdg. He inquired whether or not the cutting tool would continue to run during positioning intervals. Pease assured him it was his intention to keep the tool running; what he left unsaid was whether or not he was trying to find a way to continue metal cutting along *three axes* during repositioning intervals. His response to Burdg's question, as recorded in the minutes of the meeting, indicates he thought continuous metal cutting would be possible if the cut were a straight line in the plane of the worktable. Clearly much more system analysis remained to be done at this stage of the study.

John Parsons summed up his reactions to the October 24 through 26 meeting in a follow-up letter to Gordon Brown: "We are naturally impressed with the scope of your activities, but we are also much impressed with the enthusiasm shown for our project. . . . I have complete confidence that this project will develop in a manner that will bring pride and satisfaction to both our organizations." However, in the same letter Parsons stated that he was authorizing Snyder Tool and Engineering to proceed with a detailed design of a special purpose experimental machine because "[I] believe it to be far more adaptable to the type of work you will be engaged in than any standard machine tool you could possibly secure, primarily because it has large flat mounting surfaces . . . where you can readily mount your controls" (2–8). There was some justification for using a specially designed Snyder tool for a one-axis demonstration. Snyder made many special transfer-type machines for the automotive industry and they had a library of standard bases, slides, drives, bridges, and so forth which could be assembled in many different configurations. Hence, they could supply the project with a base and slide quite inexpensively, as a start, and later expand the equipment to three axes. Furthermore, their specially furnished machine would not be cluttered up with superfluous drive machinery that would have to be removed or adapted to the servo drives.

After the October meeting, Professor Brown, as Laboratory Director and overall supervisor of the project, met periodically with the project staff to discuss progress and future plans; occasionally he would invite Jay Forrester, as Associate Director of the Laboratory, to participate in these meetings. Forrester was well qualified to make cogent comments, particularly from the digital electronics viewpoint, and to give the numerical control people the benefit of his several years of experience as leader of the Whirlwind project. According to Forrester, he would never hesitate to offer constructive criticism if he thought it would be of help. For example, McDonough recalled that once, when he had proposed a really "wild" design for a digital data processor, Forrester informed him that his scheme was equivalent to "several Whirlwinds " and therefore unrealistic.

Evidently Forrester's frequent critiques motivated Brown to ask him to undertake an in-depth study of the project and to come up with a set of recommendations. Because of Forrester's uncanny ability to see the many interrelationships in a project, as well as his expertise in digital technology,

Alfred K. Susskind

his opinions were always valued highly by Brown. Forrester's evaluation of the numerical control project was thorough and incisive. In his report (2–9), he sized up the digitally controlled machine tool problem as being much bigger than either John Parsons or Servo Lab had initially visualized it, especially in terms of the amount of digital signal processing that would be needed. As he looked into the project, Forrester found that substantial progress had been made on a variety of possible designs for various parts of the control system, but that much more work would be necessary before a final overall design for a demonstration system could be achieved. He concluded it was much too early to build anything: much more system research would be needed, and funds beyond the $30,000 allocated to the Laboratory would be required. Forrester therefore recommended that the Parsons contract with the Air Force be amended to provide more time and funding for system research. In addition, he recommended that more manpower be added to the project, and further suggested that Pease be temporarily relieved of his teaching duties to devote full time to the project and that at least one other staff person be added to the project roster, someone with experience in large-scale digital systems design. Alfred K. Susskind, a staff member of the Whirlwind project, was the person selected to fulfill those requirements.

Jay Forrester's report was important and timely; it brought into sharp focus the fact that the Laboratory was dealing with a wholly new engineering research problem rather than one which could be solved through straightforward engineering, and that because of this both manpower and funding requirements had been grossly underestimated. However, the existence of the Forrester report posed some risk: Given John Parsons' desire for a one-axis demonstration, as expressed during the October meeting, it was quite possible that he would find the report's assessments unsatisfactory and would decide to pull back the remaining funds allocated for MIT to his own shop, where Frank Stulen continued to work independently on the problem. In actuality, this did not happen. At a January 6, 1950 meeting with Parsons, held at Servo Lab, Professor Brown expressed the strong conviction that he and his colleagues had gained enough insight into the overall system aspects of the problem to conclude that neither they nor anyone else could devise a complete operating system within the time frame and budget of the Parsons–MIT Agreement. Parsons reminded the Servo Lab staff that it had been given free reign, at the October meeting, to conduct the study in its own way, subject to the time and funding constraints of the Agreement, and expressed concern about any expansion of the project that would cause it to drag on beyond the time goals called for in his Air Force contract. He also expressed his hope that Servo Lab did not intend to "mull over" the project for a whole year. Understandably, Parsons preferred not to go back to the Air Force for more funds at a time when there was nothing tangible to show for those already spent; also, he continued to show that he was thinking in terms of a specially designed machine tool that would fabricate aircraft wing panels exclusively. Parsons obviously did not share Servo Lab's goal of trying to contrive a system that would be applicable to a broad array of aircraft parts manufacture. This last point led to a discussion of Forrester's recommendation that the economics of a digitally controlled machine should play an important role in the determination of the final system design. It

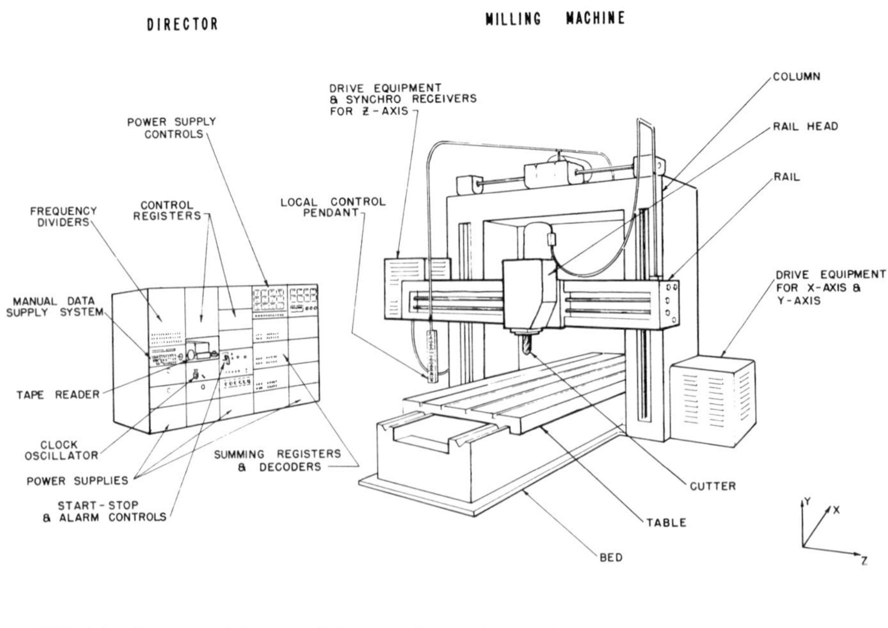

FIG. 2.1. Conceptual layout of the experimental numerically controlled milling machine.

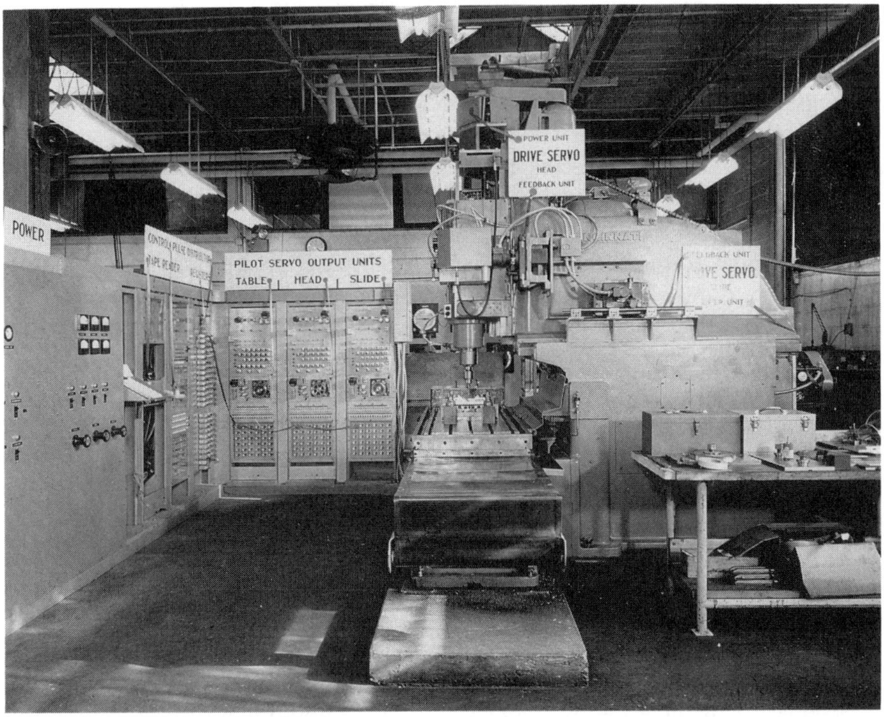

FIG. 2.2. The MIT numerically controlled milling machine.

was Forrester's contention that MIT, which was responsible for the system study, was in the best position to make economic evaluations since the economics of the automatic milling process depended heavily on equipment configuration. Parsons, once again taking a narrower view, minimized the importance of economics on grounds that there was at that time no simple way to make wing panels to the Air Force's accuracy specifications; hence by definition his original system concept was an economically viable solution. The upshot of the meeting was that further meetings would be held between Servo Lab staff and representatives of the Parsons Corporation in an effort to arrive at a mutual understanding of what the overall system should be.

As it turned out, three conferences were held, all of them in the month of January 1950. By the end of that month, the conferees had set out three objectives: (1) the development of a bench test unit sufficient to demonstrate and test the principles involved in controlling a lead screw from digital data; (2) the modification of a standard three-axis milling machine to demonstrate data control of one axis; and (3) the further modification of the aforesaid machine to provide data control over two axes. An important conclusion reached by all involved was that while such a program could produce an experimental machine that would have most of the features desirable in a fully operational machine, it might still contain constraints that would limit its use for the the contouring of wing panels.*

A revised timetable and budget also were agreed upon during the January meetings. In general terms, it was concluded that a program which would make most economical use of time and funds was preferable to a short-range crash program. In terms of specifics, it was estimated that the three objectives outlined above could be achieved within two years at an estimated cost of $200,000.† Thus, the common understanding of program objectives satisfied John Parsons' desire for a simple, bench-type demonstration of a digitally controllable lead screw and also served to bring about a mutual agreement that the project leader would think in terms of a commercial milling machine rather than a specially designed tool.

The rest of the project's first year, February through June 1950, was devoted to completion of the system study. A comprehensive report describing the features of the proposed control system was issued June 30, 1950 (2-10). This report, entitled "Provisional Specifications for the Development of a Numerically Controlled Milling Machine," set forth specifications and performance requirements for all the essential controller subsystems, packaged as shown in Fig. 2.1. Except for the machine tool itself, which remained to be chosen, the overall system illustrated in this figure bears a close resemblance to Fig. 2.2, the final embodiment of the demonstration system.

It is important to note that the numerically controlled machine tool system described in the June 30, 1950 report was fundamentally different from that originally envisioned. It is also important to understand why it was necessary to make a major departure from the original concept of an absolute coordinate system in which each positional element of the machine is specified by a different binary number referenced to a starting point. Since the goal was an experimental system that could meet commercially acceptable cutting speeds and accuracies, the design study had proceeded on the basis of a positional resolution of 0.0005 inch and a maximum cutting speed of 15 inches per minute. These numbers yielded a binary number rate of 500 numbers per second. If one punched card contained one positional instruction, the cards would have to be read at a rate of 500 per second—an impossible rate for any card or tape reader then available. Furthermore, the number of binary bits per axis that would have to be processed was quite large. For example, the total table travel of the Hydro-Tel milling machine used as the experimental vehicle was 60 inches. Hence, 120,000 different binary numbers would be required to specify all table positions within the 60 inches ($60 \div 0.0005 = 120{,}000$),and a unique 17-place binary number would

*The real clincher for using a standard milling machine instead of a specially designed machine was that the Air Force could furnish the project with a standard machine at no cost; thus the $30,000 in funds earmarked for a special machine could be diverted to other parts of the budget. Not all of the $30,000 was saved, however; the milling machine the Air Force furnished arrived damaged. It also had to be reconditioned and modified for project use.

†The Laboratory's experimental milling machine, digitally controllable along *three* axes rather than two, was completed in March of 1952, just two months beyond the estimated date.

be required to specify each of those 120,000 positions (2^{17}= 131,072). These quantities equate to 120,000 times 17, or 2,040,000 binary bits of information for the 60 inches of table motion. Another 884,000 bits would be needed for the maximum 26 inches of cross-slide motion, and a third number of similar magnitude would be required for cutter spindle motion. It was estimated that use of a unique binary number to define each 0.0005-inch resolution element would consume the entire capacity of the Whirlwind computer in order to process that data! Whirlwind I was a big machine—it had some 6000 vacuum tubes, initially, and took up the whole floor of a building. Clearly, using a computer of that size to control a three-axis milling machine was unrealistic.*

As the Servo Lab study progressed, the project staff continued to seek a way to circumvent the absolute-number system of specifying position. In so doing, the group came up with an incremental-coordinate and rate-control approach that brought the data processing problem down to manageable size. In this approach a curved surface is described by a series of straight-line cuts that approximate the curve, and commands for each new straight-line segment are given with respect to the end point of the preceding segment rather than with reference to the original starting point. The number of straight-line segments into which a curve is divided is determined by its sharpness; the gentler the curve, the fewer are the number of cuts needed to preserve the specified cutting accuracy and the fewer the commands for guiding the machine tool. Another feature of the approach is that the distance of the cut along each principal axis and the rate of motion along each axis are determined electronically by means of a train of pulses. For the experimental machine tool, each pulse represented 0.0005-inch of motion along an axis, and a unique pulse recurrence rate for the axis specified the rate of motion in the axial direction. Three sets of pulse trains were required, since in general each axis required its own speed rate in order that travel along all three axes would be properly coordinated at all points of motion and completed simultaneously. In the incremental approach, a different 17-bit number is required only for each *new motion* along a principal axis, not for each 0.0005 inch of travel. Hence, only 51 bits are required to specify one straight-line cut in three-dimensional space regardless of the length of the cut.

As an illustration of the process, assume for simplicity that motion is along only one principal axis, the axis of table motion. For a machine with a maximum table displacement of 60 inches, a 17-bit number is needed to specify a cut of that length to an accuracy of 0.0005 inch. However, once the number is set and a train of timing pulses recurring at a rate corre-

* The severity of the information-handling problem associated with a three-dimensional point-by-point positioning system was recognized very early in the MIT system study by both MIT staff and Mr Marsh of the Parsons Corporation, and much thought was given to techniques that might achieve desired accuracies with a reduced amount of information processing. In one scheme examined by Marsh, for example, cutting would proceed continuously between points specified 0.03-inch apart at rates that would not be closely controlled at all times—velocity would be checked every 0.05 inch.

sponding to the desired table speed is established, it is then necessary only to count the pulses and stop cutter motion when the count reaches the value specified by the 17-bit number. If another cut is to be made, a new 17-bit number, corresponding to the new incremental distance, is specified for the length of the new cut and the process is repeated, with the end point of the preceding cut as the starting point of the new cut.

The design challenge here, of course, was to configure a digital data processor that would enable the machine tool to execute a series of straight-line cuts in three-dimensional space without interruptions in motion. We refer readers who wish more technical details on how this was accomplished to Chapter 3; we only note here that the June 30, 1950 report on provisional specifications for the demonstration machine tool went beyond block diagrams—design details were included. For example, the controller was divided into twenty-five functional components and performance requirements for each component were established. The design study concluded that the overall system should be controlled by a master timing-pulse oscillator, in other words by a *master clock*. The clock design had been worked through in sufficient detail to specify its frequency range, stability, output pulse amplitudes and shape, voltage supply requirements, and the complement of vacuum tubes that would be included in its implementation. Similar detail was included in the provisional specifications for the other 24 components. It was estimated in the report that the digital controller would require a total of at least 171 vacuum tubes, plus those in the power supplies. When finished, the operational controller and its associated power supplies used 292 tubes.

The subject of vacuum tubes is worth a few comments here as an item of historic interest. Because of Susskind's prior association with the Whirlwind project, he was sensitive to the need to exercise extreme care in the choice of vacuum tube types for the electronic circuitry. The Whirlwind staff, in what turned out to be a major contribution to vacuum-tube technology, had discovered that ordinary vacuum tubes were not satisfactory for digital computer applications, where there was a stringent requirement for thousands of hours of on/off operation without significant deterioration in plate current during the "on" state. Several potentially useful tube types were experiencing a gradual fall-off in plate current because of a chemical reaction between their cathode cores and their coatings. This reaction resulted in a build-up of a resistance interface between the core and the coating, and the presence of this resistance in the plate-current path caused erratic operation of the digital circuits (2–11). However, as vacuum tube manufacturers foresaw an expanding market for highly reliable long-life tubes, they began to bring out products especially fabricated for digital applications. Hence, during the first several months of Susskind's association with the project, he devoted a substantial fraction of his time searching for tubes that would be be most suitable for the controller.

This effort paid off handsomely later on, during the first public demonstration of the system, when he was bombarded with questions about system reliability and the "down time" being experienced because of vacuum tube failures. Susskind was well prepared to quote impressive reliability figures. During the immediate postwar era, systems employing vacuum tube technology were considered suspect in terms of reliability and precision by equipment suppliers and users who had not worked closely with digital electronic technology. Susskind had to overcome these prejudices and was well prepared to do so.

The digital controller design that emerged from the study phase of the project was in fact the system finally implemented during the following two years. We note in passing that the conceptualization of how the controller should be put together was realized within exactly one year from the starting date of the project, with an expenditure of funds amounting to $54,000. The June 30 study report fulfilled the phase one requirement of the Parsons–MIT Agreement.

Before considering the second and third years of the numerical control project, we pause for a short anecdotal note on how the term "numerical control" was coined. During the latter part of 1950, the expression "numerically controlled milling machine" began to show up in reports and memoranda written by staff members of both the Parsons Corporation and Servo Lab, but *numerical control* as a generic descriptor for digital control in general did not appear until a year or so later, and came about in the following way. Pease and Parsons thought that positive steps should be taken to come up with a suitable term. They discussed this need over lunch on December 4, 1950, with the result that Parsons decided to sponsor a contest in order to "secure a suitable name for our digital data machine control unit." In a December 28, 1950 letter to Pease, Parsons outlined the ground rules of the contest: ". . . short words [are] preferred," the term should be a nonstandard word, "so that it may be copyrighted," and only "MIT personnel working on the project are eligible." All entries would become the property of the Parsons Corporation. The judges would be Dr Gordon Brown, R. Allen Price (Parsons Corporation), and John T. Parsons, and their decision would be final. A prize of $50 would go to the winner, whose name would be announced on or before March 15, 1951. Our records show little more except that Pease relayed the announcement to his Laboratory personnel and "numerical control" was the winning entry. According to Pease, it was he, in conference with Jim McDonough, who had submitted the winner, but whether or not they ever received the $50 and if they did, how they split it (and spent it), remains unrecorded history.

The Second Year

During the second year of the project, dating from July 1, 1950, substantial technical progress was made toward implementation of the digitally controlled machine tool system that had emerged on paper from the design study work of the first year. The technical team was now in place, with a total of nine people assigned to the project. McDonough, with the help of junior staff members, took charge of the power servo designs and the paper tape preparation unit. Herbert P. Grossimon, a newcomer to the project and the Laboratory, undertook the design of the relay logic under McDonough's guidance, while Susskind and his three assistants tackled the electronic circuit designs. Pease continued to head up both the technical and administrative sides of the project, and Robert Marsh, Parsons Corporation's on-site representative, continued to handle a variety of technical assignments.

A major issue that had to be settled as quickly as possible was the machine tool that would be used as the demonstration machine. Earlier that year a common understanding had been reached between Parsons and the Laboratory that the design would utilize a commercial size tool, but whether it would be specially designed, purchased off the shelf, or furnished by the Air Force from its stockpile was still an open question. The matter was finally resolved during a meeting held at MIT on August 5 and 6, 1950, among representatives of Servo Lab and Parsons Corporation and the Air Force's Harold Sennett. The decision was that the demonstration machine would be a standard, rather than a specially designed, milling machine, and that if a suitable machine could be located in the Air Force's inventory, it would be bailed out to the project. Robert Marsh was assigned the job of visiting several machine tool builders to study the suitability of their machines for the project. The criteria for a final selection included machine tool size, the amount of work that would be required to adapt the tool to digital control, the rigidity of the structure, and the availability of convenient space to mount the three power drives. After visiting five machine tool manufacturers, Marsh concluded that the project would be best served by choosing Cincinnati Milling's 24"× 60" vertical Hydro-Tel milling machine. The Servo Lab staff, John Parsons and Harold Sennett concurred in this choice, and a specific machine was requisitioned from the Air Force surplus stockpile at storage site AAP No.6 in Marietta, Georgia. The machine arrived (damaged) at MIT on October 25, 1950, and an outside contractor was immediately engaged to make the repairs and modifications deemed necessary by Servo Lab staff.

Meanwhile, the system implementation process moved along in accordance with the style Pease had outlined at the beginning of the project. A total of 27 subsystems (called components in project reports) was identified, and as noted above, each senior staff engineer was given responsibility for three or four of these subsystems. Every circuit in the system was first designed on paper, then the critical elements included in that circuit were procured and individually bench tested against the performance criteria that had been established for each element. If it appeared that acceptable parts were available, the circuit was "breadboarded," tested for its ability to meet specifications, and then redesigned, if necessary, until the designer was satisfied the unit would perform as expected. When the design was final, an engineering drawing was made and an engineering memorandum was written describing the subsystem's operation.

A substantial evaluation effort was sometimes required before certain circuit elements could be chosen. A case in point concerns germanium diodes, which were seen as attractive compared to vacuum tube diodes because of their small size and high efficiency. Because at that time solid-state diodes represented a relatively new technology, it was found through bench testing that only a small percentage in a batch of a hundred or so could qualify for the machine tool application. Fortunately, through exhaustive testing, four different diode types obtainable from three different suppliers could be declared qualified.

Selection of suitable vacuum tubes was a time-consuming job. As a precaution against premature vacuum tube failure, a policy of "pre-burning" tubes was established: all tubes were operated for two hundred hours *before* being tested for their ability to meet specifications, the theory being that early tube failures would show up during the pre-burn period and those tubes that survived would be likely to continue to perform reliably over an extended period of time. To speed up the pre-burn process, a test rig was constructed to permit simultaneous pre-burning of fifty tubes of a kind. Clearly, this policy contributed markedly to the reliability of the demonstration system. With the development of transistors and solid-state integrated circuitry, of course, vacuum tubes faded into the historical past.

Although the technical program advanced smoothly during the second year along well-thought-out lines, administrative matters were in turmoil. In fact, at one point shortly after the turn of the new year, 1951, the project was on the verge of collapse for contractual reasons. When the decision was reached to use an Air Force surplus milling machine as the demonstration vehicle, Harold Sennett at first thought it would be necessary to issue a supplement to the Parsons–Air Force contract before the machine could be bailed out to Servo Lab. However, after some searching, Sennett found that although a contract change was indeed necessary, he could ship the machine while the revision was in process. Another reason why a contract modification was necessary resulted from changes in time scale and funding agreed upon between Parsons and Servo Lab back in January of 1950—Parsons' contract with the Air Force (due to expire shortly) was no longer in synch with the current project development plan. Thus, in the early autumn of 1950 it became necessary to begin work on a major contract revision. While the release the previous June of Servo Lab's document setting forth system study results had perked up the Air Force's enthusiasm for the project, the availability of this written document also had provided a wider exposure for the numerical control project within Air Force ranks, thereby opening opportunities for both positive and negative reaction. The consensus at Wright Field that fall was that tangible progress in the form of a minimal operating system should be demonstrable by the spring of 1951. In addition, they asked that three proposals, for three levels of time and associated dollar amounts required to complete the project be submitted by Parsons for Air Force consideration. Pease had these numbers ready for submission to the Parsons Corporation by the end of September, 1950. His plan called for the completion of a three-axis demonstration machine tool system by December, 1951 at an estimated cost of $298,000—$38,000 more than the amount that had been estimated the previous January. Also drafted was a crash program that would complete the work four months earlier (September, 1951) at an estimated cost of $336,000, and a stretched-out version that would delay completion of the project until well into 1952. Pease also drew up detailed drafts of contract supplements for each of the three plans for John Parsons' use in his negotiations with the contract officers at Wright Field.

It became evident during a meeting at Wright Field on December 15, 1950 that contract renegotiations would not be easy. So many changes needed to be made in the existing document that the Air Force indicated it would prefer to close out that contract and start over, a process that would take several months. Also, the extra funds being sought by the Parsons Corporation—$496,000 versus the original $200,000—began to raise questions at top Air Force levels about the degree of interest within the industrial sector in the whole concept of numerical control. In fact, the Air Force negotiators decided to condition any new contract on a "reselling" of the program to its Industrial Mobilization Division—in particular to Colonel Paul Brueckner—and on the production of tangible evidence that the industrial sector did indeed have a genuine interest in the project.

During the next three weeks, both Pease and Parsons put forth a major effort to compile all the information requested by Wright Field. As part of the reselling job, Parsons prepared a detailed chronological history of the project from its inception, including reasons for the changes in direction that had emerged from Servo Lab's in-depth study of system requirements (2–3). In his document, Parsons cited favorable reactions to the project he had received from representatives of the following companies: Boeing (". . . if Parsons equipment were available today, [we] would put it to work immediately"), Bell Aircraft (". . . the Parsons equipment would be of great help in the fabrication of guided-missile airfoils"), and Cincinnati Milling ("Mr Melhope [stressed] the point that it would provide an excellent method for producing accurate models for their Hydro-Tels"). Other endorsements were cited from Lockheed Aircraft, Piaseki Helicopter, Wyman-Gordon, and Pratt and Whitney.

While John Parsons worked on his documentation, Bill Pease continued to refine his estimates of the time and funds that would be required to achieve a demonstratable three-axis digitally controlled milling machine. Thus, Parsons and Pease were well prepared to do a reselling job to the Air Force during a three-day meeting held at Wright Field from January 10 through 12, 1951. Pease summarized the results of that meeting in a memorandum to Gordon Brown:

1. Wright Field is interested in the control system development for which funds are now being expended.
2. The existing contract between the Air Force and the Parsons Corporation is in such a state that no straightforward procedure to amend the contract can satisfy all Government requirements.
3. Wright Field accepts the MIT estimates of costs on the basis that no better cost figures are presently available.

4. Wright Field consensus is that the fraction of the total proposed contract amount of $496,000 to be spent in the future by the Parsons Corporation itself ($129,000) is unreasonably high. Mr Brueckner stated that a fixed-price proposal by the Parsons Corporation aggregating approximately $400,000 might serve as a basis for amending the contract.
5. Mr Rumley, of the Parsons Corporation, stated that the financial position of his company is such that this project cannot carry the element of possible loss for the company. Furthermore, the company cannot invest in the project any amount substantially in excess of what is already invested. In consideration of this, the Parsons Corporation is willing to waive all profits. (2–12)

Evidently John Parsons was led to believe by the Air Force negotiators that the contract he was trying to work out might not happen. At any rate, in a postscript to the memorandum, Pease wrote:

> On January 18, 1951, Mr Rumley and Mr Parsons visited MIT to discuss the possibility of MIT's undertaking the work of the project on contract with the Air Force. Mr Sage and Dr Brown agreed that MIT is interested in continuing the work now in progress and further agreed that, subject to satisfactory contract basis, MIT would be glad to do this work directly for the Air Force. A meeting to discuss this proposal with the Wright Field personnel was scheduled for Monday, January 22, 1951. (2–12)

Since a new contract had to be negotiated, the Air Force was considering all its options, including that of placing the new contract directly with Servo Lab rather than with the Parsons Corporation. Pease recalled receiving a telephone request from Wright Field for information about the status of the project and for an estimate of the funds MIT would require to complete it. Subsequently, Pease prepared a document that amounted to an MIT proposal and transmitted it to Wright Field the day before Christmas, 1950. Pease informed Parsons of this turn of events, and the whole matter was aired at Wright Field on the scheduled meeting day, with the result that on the following day, the Parsons Corporation formally notified MIT that it was terminating its Memorandum of Agreement effective January 31, 1951 (2–13). In a January 23, 1951 letter to Nathaniel Sage, director of MIT's Division of Industrial Cooperation, John Parsons indicated that in recent negotiations with the Air Force, the government had "evidenced the desire to eliminate Parsons Corporation from the development phase of the project." He also expressed the feeling that a way would be found for MIT to continue the work, and he further stated that "we have enjoyed working with members of your staff and cherish the hope and belief that they will be successful in completing this project so that industry may receive the benefits we believe possible with such a system." Thus, the MIT–Parsons relationship ended, except for patent matters, discussed in Chapter 3.

At this point, it is appropriate to ask the question: Why did the Air Force not renew the Parsons Corporation contract and decide instead to work directly with Servo Lab? The Air Force personnel who were in charge at that time, Colonel Paul Brueckner, Thomas G. Edwards and Harold Sennett, are the ones who could provide an authoritative answer, but unfortunately they are either deceased or their whereabouts are not known. However, a critical review of the files, as well as interviews and conversations with several of the persons involved at the time, have led to the author's hypothesis that the Air Force's decision was based upon a set of circumstances that arose after the Parsons Corporation received its initial contract, along the following lines. As a result of the system study undertaken very early in the project by Servo Lab and Parsons, there emerged a concept of a computer-controlled machine tool which differed substantially from the concept originally proposed by John Parsons. Furthermore, it was evident that much more time and money than was originally envisioned would be required to produce a demonstration system based on the Servo Lab concept, and essentially all of the effort would be expended by Servo Lab. Pease's documentation of the Wright Field meeting on January 10–12, 1951, indicates that the Air Force appeared reluctant to provide all the funds Servo Lab needed to continue the development work *plus* the substantial amount sought by Parsons for its administration and supervision of the new contract and whatever independent development work it wished to undertake.

As stated by E. C. Rumley of the Parsons Corporation at the contract renegotiation meeting, his company was in no financial position to supplement Air Force funds with its own capital. Underlying this position was the fact that Parsons was having financial problems at that time and its creditors were overseeing the company's expenditures. Since the Air Force was satisfied that the aircraft industry was interested in the numerical control project along Servo Lab lines, they concluded that it should go ahead and that the most efficient way for that to happen would be through a direct contract with MIT. There is also evidence that the Air Force took the first steps in that direction. Pease recalls puzzlement over a call he received from Colonel Brueckner about MIT's interest in working directly with them; in Pease's mind he was already working for the Air Force through Parsons. The position Pease was being put in was awkward, and indeed embarrassing, since his relationship with John Parsons was quite amicable. This complication was recognized by Colonel Wilbur R. Carter in his 1969 interview with the author. Carter, who held important posts at the Air Materiel Command, WPAFB, and at the Pentagon, volunteered the following comment:

> I think . . . it may be very interesting for you to bring out the position of MIT, as I recall it, when they came in to talk relative to taking on the contract. They were not seeking the contract but they would take it on to go ahead and further move along the general idea of numerical control. So there was just a bit of awkwardness at that time because there was an existing contract with Parsons, and MIT was exhibiting a willingness to take it on, but not soliciting in getting the contract. (2–14)

There is also support for the hypothesis that the Parsons Corporation, due to its shaky financial condition, supported the Air Force's desire to transfer the contract to Servo Lab. In the author's 1969 interview with William M. Webster, the Air Force's project engineer on our project for nearly seventeen years, Webster stated:

> At the time I arrived [on civil service at Wright Field], the Parsons Corporation . . . was running into financial problems of sorts and they were interested in the transfer of responsibility of the contract, apparently to MIT. And in any event, when I arrived, it [the Air Force] was in the process of reorienting the project, releasing the Parsons Corporation from primary responsibility, and the establishment of MIT as the prime contractor.' (2–15)

Thus, the archival record reveals a consensus among the three parties that the Parsons Corporation should withdraw in favor of Servo Lab, a conclusion that resulted in another major turning point in the project, now placed in an uncertain position through this new turn of events. Indeed, the project nearly came to an end: from February 1, 1951 MIT would be without contractual support, and there was no guarantee that support would soon be forthcoming. While Pease was sure the Air Force wanted the project to go forward, he was also well aware of the time it would take to get the necessary paperwork past the various sign-off levels of the bureaucracy. If the work was to continue uninterrupted, it would be necessary to persuade top-level Institute officials to bankroll the effort pending receipt of an Air Force contract. The job of making this decision fell to Nat Sage, who lost no time in contacting Wright Field. Early in February, Sage wrote to Pease about his conversations with J. B. Straley, the contracting officer for the Air Force's Manufacturing and Methods Branch and Major William Adams, Branch Chief (2–16). Sage reported that his conversation with Adams was pleasant and that Adams was apologetic that MIT had not been informed that it might be very difficult to get the $200,000 it was seeking. However, the residual amount of $60,000 remaining in the Parsons contract, then in the process of being de-obligated, might easily be transferred. Sage went on to say that he had explained his difficulty to Adams, which, as Sage expressed it, was "to find out whether we were working for money or for love," and that he needed assurance that the Institute would be covered from February 1, 1951 on, an assurance Adams refused to give him. Adams then suggested that the project "slow down," a clue to Sage that there was a genuine desire at Wright Field to continue the work. However, Sage was not in a position to accept a slowdown plan on the grounds that if the Laboratory staff were to turn to other research, it would not be possible to regain their services at a later time. Sage also learned that Straley was "not constitutionally opposed to writing in a clause for costs in anticipation of the contract," but the Wright Field people who wanted the work would have to be the ones to initiate the action. Thus, Sage was not able to extract a guarantee from the Air Force that eventually funds would be forthcoming, with a February 1 starting date. None the less, given his many years of experience in negotiating government contracts, Sage had evidently heard enough to convince him that a contract would eventually arrive. Hence, he authorized the work to go forward. The pace, however, had to be curtailed simply because it was impossible, due to the ongoing Korean conflict, to purchase certain critical items of equipment without a Defense Order (DO), and DOs could be obtained from the government only when the requisitioner had a government contract in force. Because of this inability to purchase equipment, we estimate that the project was set back at least two months.

Sage's judgment that the Air Force intended to continue the project proved to be sound. Toward the end of February, Pease was able to report to Brown that he had received a TWX communication from Wright Field indicating that funds for a new contract had been made available, approval had been granted by the Production Resources Division, and the procurement process was expected to begin within the week. Nevertheless, the wheels of progress continued to turn ever so slowly in the eyes of the project staff members. Finally, Albert F. Sise, Servo Lab's executive officer, was able to report to the MIT administration in early April that: "Mr Straley has finally produced the long-awaited contract number—AF33 (038)-24007," and that a DO number had also been issued. Although the two-month time period required to write the new contract seemed interminably long to project members, in retrospect the Air Force did a remarkably quick job of pushing through the paperwork. Obviously they wanted the work to continue, and there was clear recognition of the financial circumstances under which the Institute, a nonprofit organization, must conduct its research projects. So, effective April 3, 1951, the technical work on numerical control resumed at its normal rate.

Changes in Leadership

Having successfully completed negotiations for the new contract, thus ensuring financial stability for a while longer, Pease agreed to take on a new, temporary assignment as a staff member of Project Charles, a government-sponsored Institute study to look into issues relating to defense against intrusions of low-flying aircraft.* Replacing Pease as project engineer was James O. McDonough, Pease's senior staff engineer. Pease continued to be responsible for contract matters, but McDonough now took day-to-day charge of the effort and continued to do so for the next four years. Pease returned to the Laboratory early in 1952, but in a new capacity: he was appointed Laboratory Director, succeeding Professor Gordon S. Brown, who by now was Associate Head of the Electrical Engineering Department and slated to become Department Head on July 1, 1952.

The Third Year

The project's third year, beginning July 1, 1951, was one of accomplishment. It culminated in a working model of the three-axis digitally controlled milling machine. During this period, the Air Force was most accommodative of the Laboratory's ambition to complete the three-axis director

*An outgrowth of the Project Charles study was the establishment of MIT's Lincoln Laboratory. Its mission was, and still is, research in support of the national defense needs of the country.

system as expeditiously as possible. For example, Pease was able to report that, as a result of phone conversations with Colonel Brueckner and Tom Edwards (our new Air Force project engineer), ". . . the famous first-axis demonstration is off. The Wright Field people see no reason why we should not continue on toward our three-axis machine." As a result, implementation of the entire machine tool system could go forward without the need for Air Force approval on an axis-by-axis basis. Actually, one axis was implemented and tested first, then the second and third axes were completed.

As time went on, there was an increasing awareness in industrial, government and academic circles that MIT was engaged in an application research program involving a rather unorthodox real-time digital control system. This awareness undoubtedly came about because of industry's contacts at Wright Field and through direct contacts between Servo Lab personnel and machine tool users and suppliers. In early 1951, staff members began accepting invitations to present papers at professional society meetings, and these in turn prompted additional curiosity about what was going on at the Laboratory. Then too, the "open door" policy which the project maintained helped to propagate information about numerical control. Thus, Lab records show that during the ten-month period from August of 1951 through May of 1952, a total of 250 people from 101 different domestic and international organizations—industrial, governmental and educational—visited Servo Lab to discuss the work and see the system at its various stages of completion. It was becoming clear that the research was attracting interest, not only in the possibilities for numerical control of machine tools but also in real-time digital control in a broad sense.

The experimental machine tool system was up and running by March of 1952, two years and nine months after the project started. Figure 2.2 shows the installation. The next step was to gain experience in operating the machine and to gather data about its accuracy and reliability. In order to put some realism into our evaluation, the staff obtained from the aircraft and machine tool companies drawings of a variety of parts for fabrication on the experimental system. By the end of May 1952, 503 operating hours had been logged, of which approximately half represented engineering test time. A total of 63 hours was devoted to machining parts—the first job being a radar lens for the Antenna Laboratory of the Air Force's Cambridge Research Center and the second a forward drag strut supplied by the Boeing Company (see Fig. 2.3 and Fig. 2.4). The remaining hours represented demonstrations, troubleshooting, and miscellaneous machining.

With the three-axis digitally controlled machine now a reality, the Air Force wanted the Laboratory to "unveil" it to the user and supplier communities and to prepare a comprehensive report for wide distribution. These matters were discussed at an April 22, 1952 meeting held at Servo Lab; present were Lieutenant Colonel Adam Altglass, Colonel Brueckner's replacement, and William M. Webster, who followed Tom Edwards as the Air Force's project engineer. With respect to the proposed report, a decision was made that it would be written in two parts: Part 1 to be a brief description of the machine from a business point of view and Part 2 a detailed technical description. Part 1 was issued in midsummer 1952 and Part 2 was released a year later (2–17).

It was also agreed that the official demonstration of the machine should be held sooner rather than later, but it would be more effective if pertinent economic data could be compiled. The plan devised for obtaining this econ-

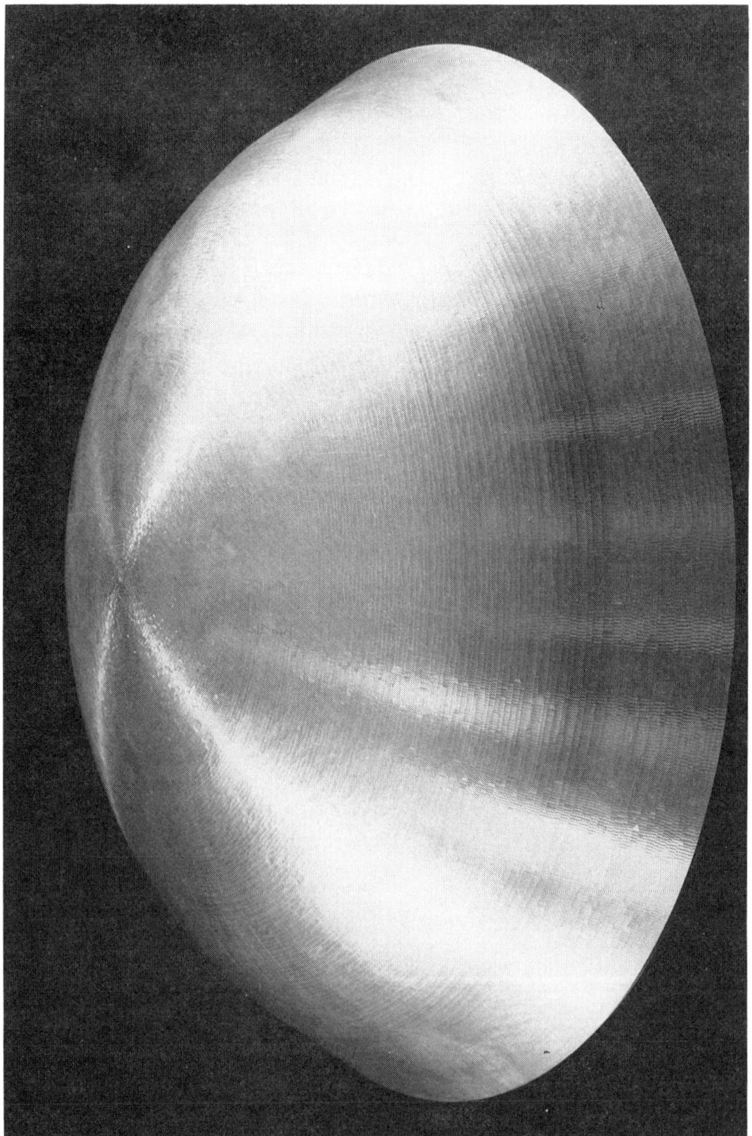

FIG. 2.3. The first piece machined on the experimental milling machine, a subassembly for the male section of a Luneberg lens.

omic data was straightforward: drawings for representative aircraft parts would be solicited from airplane builders; information needed to program the director would be derived from the drawings, and the whole process of programming and machining would be carried out under closely controlled conditions.

It was during the latter part of June 1952, on one of McDonough and Susskind's many trips to solicit aircraft parts from companies across the country, that a firm date was set for the official demonstration. One stop on the itinerary was Wright Field, for a meeting with the sponsor in the persons of Bill Webster, our project engineer, and Colonel Altglas, his immediate superior. It was here that word was received that Colonel Wilbur R. Carter, then Chief of the Industrial Services Branch of the Production and Resources Division of the Air Materiel Command,would like to have a public demonstration of the experimental milling machine for the aircraft industry as soon as possible. In the ensuing discussion of timing and dates, a plausible argument was made for delaying the demonstration on the grounds that some firm economic data would be in hand. In the end, however, Colonel Carter's desire for an early presentation prevailed, and the dates were set as September 15, 16 and 17, 1952, less than three months away. In the following weeks work proceeded at a feverish pace on all fronts—including programming, experimental machining and economic analysis—as preparations for the demonstration fell into place.

The public demonstration was an important milestone in the evolution of numerical control. A total of 242 representatives from 130 industrial organizations and government agencies got a firsthand look at NC in action,and were given ample opportunities to exchange views with Servo Lab staff and among themselves. The announcement of the demonstration, a list of the

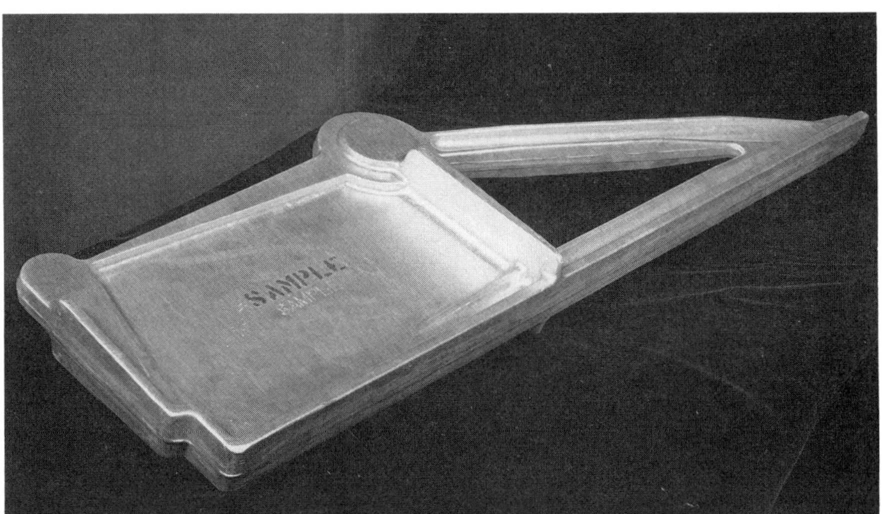

FIG. 2.4. Forward drag strut, an aluminum forging requiring contouring, facing, and slotting.

participating companies, and the number of representatives from each company appear in Appendix B. The first day of the three-day session was directed toward the machine tool builders, the second day was oriented toward those interested in more general applications of the technology, and the third day was primarily for manufacturers of aircraft parts. Each day's format was identical: a morning session on numerical control and features of the experimental NC milling machine and an afternoon session consisting of a three-part round-robin program on tape preparation, programming, and a machine tool demonstration, followed by a question and answer period for the entire group.

The question and answer sessions held at the end of each day were particularly revealing (2–18). The participants sought information on a broad range of topics. Some wanted technical details about the electronic and mechanical equipment, others raised questions about the reliability of the electronic director and who would maintain the electronics in an operational environment. When Susskind pointed out that his log showed that it took only ten minutes to trace a system failure to a faulty vacuum tube, one aircraft company representative commented: "Of course, we don't have the staff of MIT [at our disposal]."

On the reliability issue, Susskind was able to quote from his carefully kept records: "The director has been in operation now for somewhat over 4000 hours and in our records the director has been under trouble . . . two percent of that total time." For a newly designed system, composed of 292 vacuum tubes, this 2 percent downtime was an impressive figure. No one could anticipate, of course, the dramatic further improvement in reliability that would result from the displacement of vacuum-tube technology by present-day solid-state integrated circuitry.

As anticipated, economics was another subject on the minds of the participants. There was serious concern that the elaborateness of the digital director and servos would up the cost of the overall machine tool and control system to an amount that would make the whole approach financially unattractive. Lack of experience in costing out systems of this kind prevented the Servo Lab staff from nailing down cost figures precisely. However, D. C. Hierath of Hughes Aircraft made a cogent observation:

> I think you fellows are missing one of the best points of the whole thing, and that is: Here you've been talking about dollars for a machine with a control. It seems to me one of the real points [is] that with one control you can gang up a whole flock of machines, and they eliminate a whole flock of personal operations. (2–18)

Susskind quickly seized upon this point, and added:

> And, what's more, you can readily conceive of a plug-in director which, in the morning you use to guide a milling machine, and in the afternoon you use to guide a grinder, and for the next shift to run a lathe. This is entirely . . . practical because the director surely doesn't know what's going on at the machine end. All it knows is that it's cranking shafts. (2–18)

One might reasonably ask if this bit of dialog was a prediction of what we today call *flexible manufacturing*, in which an assemblage of interchangeable machine tools, interconnected by automatic materials-handling equipment, is operated under the control of one central digital computer.

The demonstration of the three-axis digitally controlled experimental milling machine brought to a successful close the first phase of the project, three years and three months after its inception. The total expenditure of funds by MIT in this endeavor over that time span was $360,000, of which $280,000 was billed directly to the Air Force and approximately $80,000 was billed indirectly through the Parsons Corporation.

The experimental machine-tool system did not go beyond the proof-of-feasibility stage. It was developed during a period of rapid advancements in the field of electronics; within a few years it was possible to replace the relay storage it used with magnetic-core storage, and vacuum tubes with transistors. As a result, the machine-tool builders went their individual ways and implemented their own designs with the most recent technologies available.

Nevertheless, the experimental system remained in the Laboratory for 15 years after the first public demonstration, during which time it was a showpiece for visitors to the Institute from all parts of the world. In 1959 we offered the complete system, including the digital director, to the Smithsonian Institution for an exhibition in its Hall of Working Tools, but after much deliberation the Smithsonian decided against taking it on grounds that a full-time person would be needed to operate it and another would be required to maintain it.

The system was finally put up for bid in 1962 as surplus property by the government. The electronics were purchased by a private individual and the machine tool went to Cincinnati Milling Machine Company where it was reconditioned and resold.*

*The Cincinnati Milling Machine Company is now known as Cincinnati Milacron, Inc.

3

THE EXPERIMENTAL MACHINE TOOL

Our purpose in this chapter is to provide insights into the design and operation of the Laboratory's experimental machine tool without going into minute detail about circuit designs, many of which are now obsolete because of the advent of solid-state technology. Hence, our discussions are principally at the functional block diagram level. Design details of the mechanical, hydraulic and electronic features of the system are presented in the final technical report on the machine (2–17).

Description of the Machine

The MIT numerically controlled milling machine is shown pictorially in Fig. 2.2 and in block-diagram form in Fig. 3.1. As indicated in Fig. 3.1, the system has two major parts: the director and the machine tool itself. The machine tool was a government-surplus reconditioned Cincinnati Hydro-Tel milling machine modified with servomechanism-driven lead screws so that its three basic motions could be controlled automatically by means of servomechanisms. The director was a small, custom-designed special purpose digital processor that accepted machining instructions stored on paper tape in numerical form and provided analog signals to drive the machine tool.

The system operated as follows. The workpiece was set up on the machine tool table in a conventional manner, the speed of the cutter spindle was selected, coolant flow was adjusted, a previously prepared paper tape was inserted in the tape reader, and the cutting operation was started. Thereafter, the paper tape controlled the process automatically in accordance with the information stored on the tape. This information included the sequential order of the cuts, and for each cut, the direction and traversal distance along each axis to be completed simultaneously in a specified time. Codes were also inserted to denote a temporary stop—a tool change, for example. The machine was capable of being programmed for roughing or finishing cuts, as desired, and overshoot at the end of each high-speed cut could be avoided by breaking the cut into two or more segments of successively slower-speed.

An important feature of the system was that it contoured a curved surface by using a series of short straight-line cuts to approximate the curve—the sharper the curve, the more straight-line motions needed to ensure surface

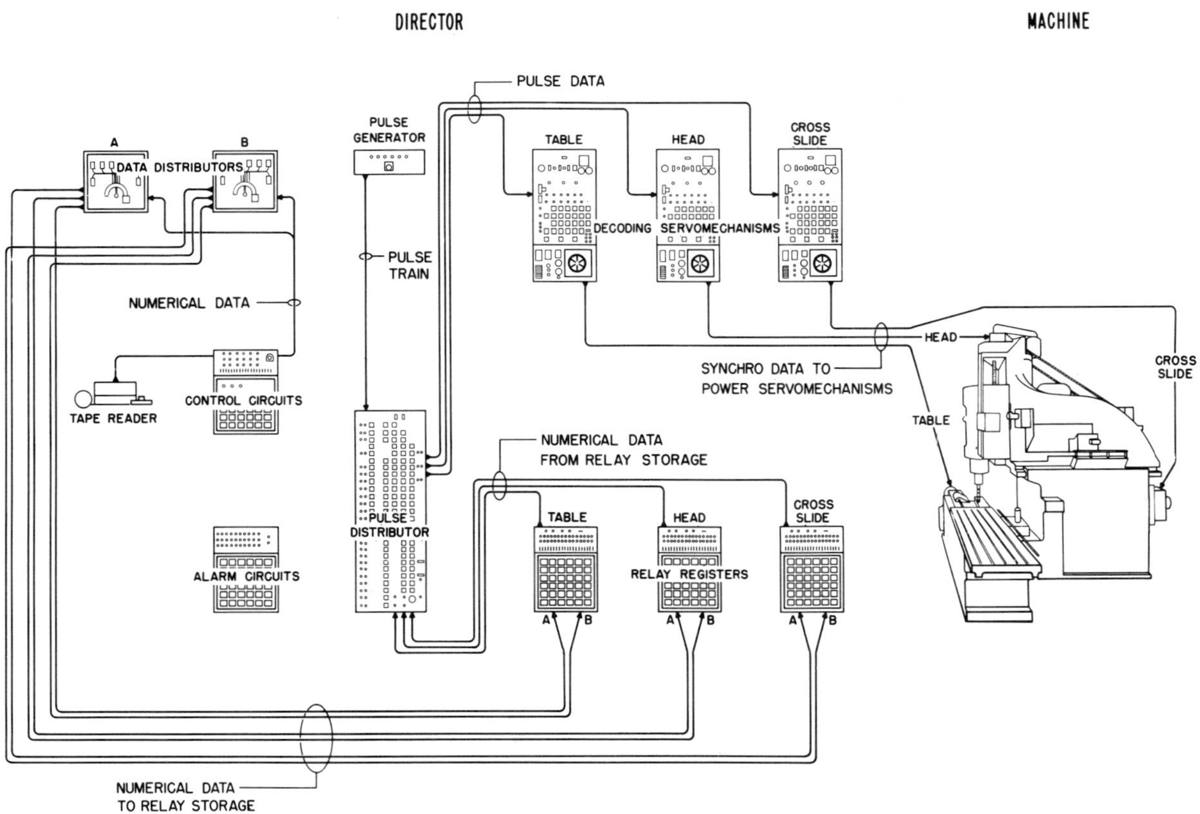

FIG. 3.1. Numerically controlled milling machine functional block diagram.

accuracy and desired smoothness. Three sets of servomechanisms were included in the design, one set for each axis, and each set consisted of an instrument servo and a power servo. Each instrument servo received trains of pulses from a pulse distributor and converted them into shaft positions at low-power level, the power being only that necessary to drive a synchro-transformer. The synchro-transformer output signals served as the input to its corresponding power servo, the purpose of which was to amplify the input to the power level required to drive the machine tool slides. The system was designed so that once a cutting operation started, cutting would continue without interruption until a different setup for the workpiece was required or the cutting tool had to be replaced.

The data input subsystem is shown in Fig. 3.2. A pair of information storage registers was provided for each axis, and designated the A and B registers. While a block of instructions in the A register was being executed, the next block was read into the B register. Then, upon completion of A-register instructions, the B register would immediately take over, thus allowing A to be loaded with the next block. In this way, machine tool control passed back and forth between the two sets of registers without interruption, and metal removal was continuous along a sequence of straight-line paths, each of arbitrary length and direction in three dimensional space.

One can think of the three pairs of storage registers—one pair per axis—as buffers between the permanent (paper tape) storage medium and the rest of the system. The tape was a standard Flexowriter seven-hole tape commonly used at that time for automated typing and also widely adopted as a data input medium in digital computers. There was space on each line for a maximum of seven holes plus a small hole, present in every line, which was engaged by the drive sprocket of the tape reader. Some of the input information, such as start-of- block signal and block length, was contained in individual lines across the narrow dimension of the tape.

The block length, which specified the number of bits in the motion lengths of a block, served a dual purpose. Permissible values varied from 10 to 17 bits,

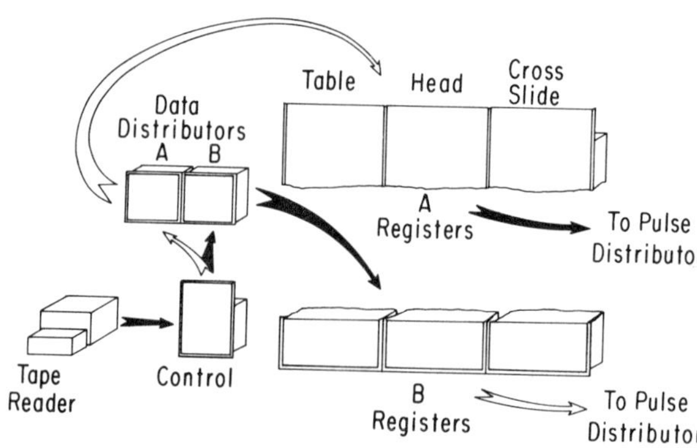

FIG. 3.2. The data input subsystem.

and the maximum motion per axis therefore varied from one-half to 60 inches per block. A saving of tape input lines was thus possible for short cuts. At the same time, the block length changed the block execution cycle of the director in powers of 2—from a cut time of 2 seconds for 10 bits to 256 seconds for 17 bits. The cutting rate was therefore almost infinitely controllable by the combination of cut length and block length. For example, a one-half inch cut made with block length 10 had a feed rate of 15 inches per minute, but the same cut with block length 17 had a feed rate of only 0.12 inch per minute.

The lengths and directions of the motions along each axis were encoded in longitudinal direction on the tape, that is, lengthwise along the tape. This information occupied from 11 to 18 lines, depending on block length.

Three of the seven spaces on each crosswise line contained one bit each of the binary encoded machining information for the three axes; the other four contained an error checking code used to detect errors during the reading of the tape. If a single error in a line was detected, the system could automatically correct it, but if two errors existed in a line, further reading of the tape would be suspended. The Hamming system of error detection codes was used (3–1).

A key feature of the director design was that information on the tape was organized into blocks, with each block (command) containing all the instructions needed for *one straight-line cut in an arbitrary direction.* Each information block specified:

1. the *distance and direction* each of the three machine slides should traverse; and
2. the *time* required to traverse the distance at the desired feed rate.

Since the commanded motions along the three axes occur simultaneously and result in cutting motion along a straight line in three-dimensional space at constant speed, a new block of information was required only when the direction or speed of motion had to be changed. Hence, a fix-speed, straight-line cut of any length could be executed with a single command. The ability to work out a successful design that coordinated the motion of multiple axes and called for a new command only when a change was required in direction of motion or in cutting speed was a major technological breakthrough. The concept was the essence of the idea of *numerical* control. It should be evident that this design approach greatly reduced the amount of information needed for a machining operation over that which would be necessary if every point along the way had to be specified to the required degree of machining precision, say, one instruction for every 0.0005 inch of motion, or 2000 instructions per inch .

Observe in Fig. 3.2 that two data distributors appear between the buffer storage registers and the controller–tape–reader combination. The data distributors consisted primarily of two stepping switches, one for each register. Their purpose was to put tape information into the proper parts of the buffer. For example, the first line of each block contained information about the time duration for that block and the second line specified the direction of motion for each axis—left or right (+ or −) in the case of table motion, in or out for the cross slide, and up or down for the cutting head. It was the job of the data distributor to connect the tape reader during the reading of the first line to the part of the buffer that stores *time* information; during the reading of the second line, to the part of the buffer that stores *sign* information, and so forth. .

The element that coordinated all the activities of the data input system was the *control unit*. It served to clear the system at the beginning of a program and feed out the blank leader of tape, cause the reading of a block upon receipt of a signal from the pulse distributor, stop the tape reader after each block was read, and alternately connect one of the two buffer registers to the pulse distributor system, the purpose of which is explained later.

In summary, information was transferred from the permanent storage medium of paper tape to temporary storage in accordance with the following steps:

- First, the system is cleared and the blank leader is disposed of.
- Second, the tape reader is started and a block of tape is read into one of the buffer registers, for example, B. During this time, the A register, which is clear (it contains only zeros), is in control of the machine and causes it to stand still.
- Third, upon receipt of a transfer pulse from the pulse distributor, control of the machine is transferred to the B register, thereby starting execution of the first command. Simultaneously, the tape reader is restarted, causing the second command (block of information) to be stored in the A register. Again, the tape reader stops as soon as the block is read.
- Fourth, upon receipt of another transfer pulse, control is transferred to A, causing the second command to be initiated without a pause. The B register is now cleared and the third command is read into B.

This to-and-fro process continues to the end of the tape, whereupon the system stops after executing the final command.

Another essential part of the director was the *data interpreting subsystem*, consisting of a master clock pulse generator operating at 512 pulses per second and a circuit called the *pulse distributor*. The latter circuit received signals from two sources, the master clock and the buffer-storage registers, and

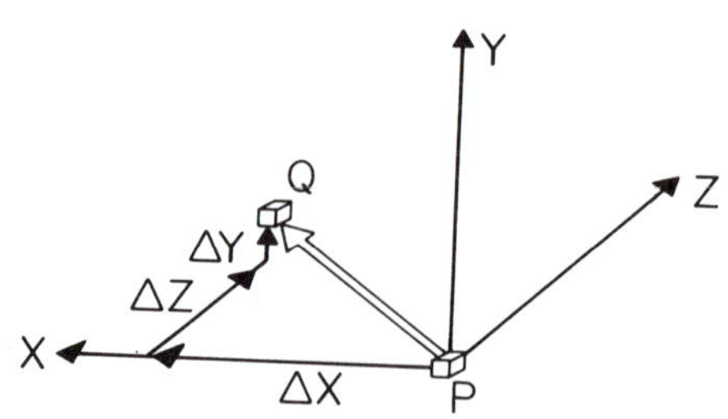

FIG. 3.3. Motions along the principal axes for a straight-line cut between points P and Q.

provided three sets of output pulses to the instrument (decoding) servomechanisms, one set for each machine axis. Each decoding servo received its signals over two separate lines, one for each direction of motion. As indicated before, a block of tape contained information describing the distance through which the machine should be moved along each principal axis and the time duration of the straight-line cut in three-dimensional vector space.

Now, in order that total tool motion in vector space be an accurate straight line, each axis should of course execute its command at constant speed and moreover the motion along each axis should be completed *simultaneously*. It follows, therefore, that in general it was necessary for the three *axial* speeds to have different values. The data interpreter subsystem determined the proper values. An example will serve to illustrate the process. As a preliminary, we note that the pulse rate of the master clock (512 pulses per second, or 30,720 pulses per minute) was chosen to meet the nominal 15-inches-per-minute cutting speed requirement. The maximum speed along each axis was then 30,720 times 0.0005 inch per pulse, or 15.36 inches per minute. Bear in mind, too, that cutting speed, whatever the value chosen, was the speed of cutting along a straight line in three dimensional space—speeds along each principal axis were a lesser amount and were determined by the required amount of motion along each principal axis and the duration of the cut. Assume, in Fig. 3.3, that motions along the three principal axes are ΔX=1.0000 inch, ΔZ=0.5000 inch, and ΔY=0.1865 inch, and that traversal of the distance between P and Q is to occur in five seconds. Two thousand pulses are needed to execute the 1.0000-inch motion in 5 seconds along the X axis (1.0000 inch equals 2000 half-thousandths and each half-thousandth inch corresponds to one pulse); 1000 pulses are needed to cause one-half inch traversal in the Z direction in the same time interval of five seconds; and for the Y axis, 372 pulses are required. It was the function of a frequency divider circuit included in the data subsystem to accept 2560 master clock pulses (512 pulses per second times five seconds) and to route the 2000, 1000 and 372 pulses to the X, Z and Y axis decoding servos, respectively.* The pulses for each servo were carried over one of two lines, depending on the algebraic sign of the specified motion. Thus, if each axis of the machine always moved exactly 0.0005 inch forward or backward upon receipt of a pulse, the center of the cutter never departed more than 0.0005 inch from the true straight line specified by the tape.

In addition to deriving the necessary frequency divisions from the master clock signal and distributing pulse data to the decoding servos, the data interpreter contained circuitry for performing other functions such as providing transfer pulses to the storage registers at the end of a straight line cutting interval, pulse checking circuitry to determine loss of pulses, and an alarm pulse in the event of pulse loss.

---

*In the example, the length of the cut in the vector direction PQ is $\sqrt{(1^2+0.5^2+0.1865^2)}$ or 1.133 inches, and the cutting speed in that direction is 1.133× 60/5, or 13.6 inches per minute.

Tape Preparation

The director of a three-axis numerically controlled milling machine can be likened to a machinist with three hands who has the skill and dexterity to control all three axes of his machine simultaneously. A digital director, however, has no inherent skills of its own; it responds only to instructions supplied to it by programmers and machinists who plan all machining operations ahead of time and store them as commands on paper tape or other storage media. Tape preparation was just one facet of a much larger task which we call job planning, a process that preceded machining and involved decisions about cutter type, size and speeds; amount of stock to be removed during each cut; number of setups of the workpiece; and the sequence of machining operations; as well as the encoding of all machine instructions in numerical form and then keypunching them onto the tape through use of a keypunching machine.

The design of jigs and fixtures for holding the workpiece in place during machining was part of job planning and followed traditional practice. There was a new challenge, however, because the great flexibility of numerical control provided a greatly expanded opportunity for ingenious scheduling of cuts to minimize the number of setups and cuts and the amount of human interaction. As discussed in the next chapter, numerical control, from its inception, showed great promise for reducing machining time, but machining efficiency came at the expense of time consumed in programming for that efficiency.

A considerable amount of mathematical calculation was required to specify machining operations. Since metal removal occurred at the edge of the cutting tool, the locus of the tool center was offset from the cutting surface and had to be calculated from knowledge of cutting-tool geometry. It was this locus that was divided into straight-line segments which were made as long as possible without departing from the desired contour by more than the allowable tolerance. Because each straight-line segment required only one instruction block, plus perhaps one or more slowdown blocks, the longer the segments, the fewer the number of instructions. For early experiments with the MIT machine, the computational and encoding steps were carried out manually, with the aid of electromechanical desk calculators. However, if work surfaces could be defined by mathematical equations, digital computers could be employed advantageously. Consequently, the Whirlwind I computer was used to program the mathematically definable surfaces of some jobs and to produce properly coded tape. When the MIT numerically controlled milling machine was demonstrated in 1952, it was already clear that the next research area needing attention was that of programming. The subject of computer programming research for numerical control is discussed at length in the next chapter.

Some Design Details

The approach that Servo Lab staff took in the development of the experimental NC machine tool merits discussion and consideration as a general model for engineering research projects. The system study undertaken at the outset of the project not only brought into clear focus an understanding of what the digital director had to do in order to meet machine tool performance specifications, but through a succession of trial design approaches in which each design represented an improvement over its predecessor, a final design was achieved which was "the best and most efficient" in the eyes of the designers. Thus as the design evolved there was a gradual realization that an enormous amount of information would have to be programmed onto the tape if every point on the cutting surface had to be specified to the required precision, and that the tape would have to be fed continuously through the reader as cutting proceeded. Hence, a concerted effort was made to create a design that would circumvent these problems. The design that was finally worked out depended on the successful implementation of several unique circuits, as well as upon their integration into an overall workable system. One such circuit was the *frequency divider*, which provided pulses for controlling cutting rates simultaneously along the three principal axes. The frequency divider concept was crucial to the final system design. It was originally devised by William Pease during the course of his system study, but as Pease later explained, the design was cumbersome and entailed an exorbitant number of vacuum tubes. In his search for a simpler approach, Pease discussed the need with his colleagues, Jay Forrester and Robert Everett. Before the session broke up, Forrester had sketched out a new design, much simpler than Pease's, which proved to be the solution that he had been seeking.

The choice of a low-power instrument servo of special design as a pulse-chain decoder followed by a power servo for the machine-axis drive is another case in point. This combination proved to be a good choice, especially since it was possible to work out an instrument servo design in which digital-to-analog signal conversion took place in the error channel, where signal magnitudes are a small fraction of the total signal handled by the servo. In addition, the dynamic characteristics of a servomechanism can be easily adjusted to provide internal smoothing (filtering) of the discrete digital signals at its input. The tandem low-power–high-power servo configuration made it possible to locate the instrument servos close to their signal sources within the director and to drive the machine-mounted power servos over low-power, low-signal-frequency lines.

The decoding servo was an ingenious piece of circuit design for its time. As shown in the simplified block diagram in Fig. 3.4, it had to accept two sets of pulse-type input signals, one for clockwise rotation and the other for counterclockwise rotation, and produce corresponding clockwise or counterclockwise rotations of the servomotor shaft. To accomplish this, output shaft rotations were digitally encoded into two sets of pulse-type signals, one for each direction, which were fed back to the input. These +/− feedback signals, together with the input pulses, were applied to a summing register, basically an up-down counter, that could count the algebraic difference be-

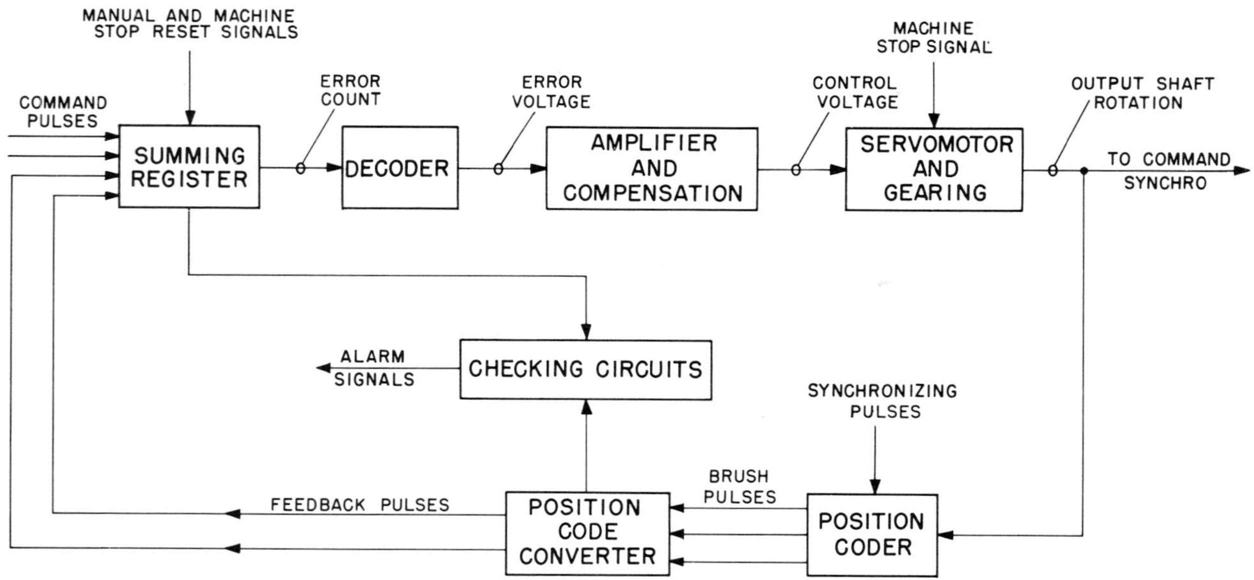

FIG. 3.4. Decoding servomechanism block diagram.

tween input pulses and feedback pulses. The algebraic difference, which represented an error, was converted to a step-wise analog voltage that served as the input voltage to an analog servo. The servo output was always in a direction to minimize the error at its input. At the time very little experience existed as background for this kind of servo design, especially in the realm of digitally encoding shaft positions (analog-to-digital converters) and digital error detectors.

There were other unique features in the overall design of the director. Although the use of relays as buffer storage registers was a conservative approach to storage, the employment of dual registers, together with high-speed electronic transfer of machine control from one register to the other, enabled the machine to operate *continuously* from data that were *discontinuous*. Error checking was also provided in critical parts of the system in order to protect the workpiece from damage. As an example, loss of master clock pulses had to be guarded against; otherwise, should the supply of such pulses be lost, the decoding servos would become unstable and drive the machine tool unpredictably.

In summary, the director represented a remarkable piece of work, not only in terms of the way it executed its function as a real-time digital controller for a machine tool, but also from the viewpoint of the variety of unique circuits that had to be contrived in order to meet performance and reliability specifications.

Patents

From the very beginning of the Parsons–MIT relationship, both parties were alert to possibilities for patents. The Memorandum of Agreement signed by Parsons and the Institute included the standard MIT patent clause in force at the time. All Institute employees working on the project were required to sign an agreement whereby they would communicate inventions to the sponsor, execute patent applications, and assign to the sponsor such applications and any patents that might be issued thereon. The Institute agreed that the sponsor would have the right to file and prosecute patent applications in the name of the inventor but with assignments to the sponsor (the Parsons Corporation), and that the sponsor would have full rights to patentable material during the first ten years from the date of filing. Other features related to the handling of uses of the invention outside the field of research. MIT had full rights, including the sole right to license others, and filing rights in the event the sponsor chose not to file. No mention of government rights was made; the Agreement was exclusively between MIT and Parsons.

Parsons Corporation retained its own patent attorneys, and during July, 1949, they conducted a search of patents that might bear on Parsons' Cardamatic machine tool idea. The attorneys found that the idea of employing punched or perforated strips and cards as means for controlling the motions of milling machines and lathes was truly an old idea. The concept closest to

Parsons' proposal was contained in US Patent No. 1,849,642, issued March 15, 1932, to one Schenker, thought to be a Swiss citizen. The Schenker patent was a completely mechanical system that employed a punched card to control motions of cutting tools—a lathe, for example. The card was divided into five vertical columns, each of which contained instructions for the controller, which could be a mechanical computer whose output was governed by the patterns of holes on the punched card. Parsons' attorneys expressed the opinion that, had not the 17-year period of the patent already expired, the Cardamatic idea would have infringed several claims in the Schenker patent. Many other patents pertaining to punched card control of machine tools and to processes outside the machine tool field also turned up during the search.

Our records show that the subject of patents invariably came up for discussion at meetings between Parsons and Servo Lab personnel, that Parsons' attorneys made several visits to the Laboratory on patent business, and that Parsons established a separate account at the Institute to deal with patent matters—MIT account DIC 7039.

An application for a patent on the machine tool system described in this chapter, and based on the configuration shown in Fig. 2.1 and Fig. 3.1, was filed August 14, 1952, and was awarded US Patent No. 3,069,608, entitled *Numerical Control Servo System*, on December 18, 1962. The inventors were Jay W. Forrester, William M. Pease, James O. McDonough, and Alfred K. Susskind; 69 claims were allowed.

The process of getting the application finalized and into the US Patent Office was an arduous one, and consumed an inordinate amount of administrative and technical staff time because of complications that arose from the transference of project sponsorship from Parsons Corporation to the Air Force prior to the filing of the application. While the Parsons agreement was in force, patent terms were governed by the rather simple, straightforward statements set forth there, but once sponsorship shifted to the Air Force, patent matters were governed by a more complex set of clauses. Then also there was the gray area in which the Parsons contract with the Air Force included the government's patent clauses, while concurrently Parsons and MIT were operating under an agreement that included different patent phraseology. Nine months after termination of the Parsons agreement, MIT's patent attorney, Melvin R. Jenney, in a November 8, 1951 memorandum, reported the status of the patent application to Servo Lab's executive officer, Albert F. Sise. The substance of this communication, based on a telephone conversation that day between Jenney and Parsons' patent attorney, was that the work of preparing the patent application was about two-thirds complete, but as yet no drawings had been started since Parsons had authorized no funds for completing the job. Some concern was expressed that the government might step in and make the filing, since Servo Lab was now operating under Air Force sponsorship. However, if preparation and filing were at government expense, then all rights would revert to the government. The possibility that MIT might undertake the task of filing was also expressed, but if it were to do so, then the Institute would in all probability seek additional rights beyond those delineated in its original agreement with Parsons.

The slow progress being made toward a filing date was cause for concern because of the necessity to file within one year of publication of patentable information. Servo Lab's DIC 6694, Engineering memorandum No. 2, containing complete design specifications for the NC milling machine, was released in June of 1950. This memorandum, however, had a restricted distribution. In addition, two articles describing the milling machine had been printed during 1951 in MIT periodicals— *Reports on Research* and *The Technology Review*. It was not clear whether these articles constituted publication, but their existence was clearly causing uneasiness over the filing delay. There was another reason why it was now important to file an application: an increasing number of inquiries was being received from industry about the patent status of various subsystem designs included in the proposed machine, particularly the decoding servomechanism. Industry evidently was awakening to the potential of servos that could be used as electronic digital-signal to mechanical analog-signal converters and for other applications.

The issue of who had rights to the invention was finally cleared up in a November 8, 1951 memorandum from Colonel William Adams to Servo Lab's James O. McDonough, in which he transmitted the ruling of the Air Force's Office of the Judge Advocate, Patents and Royalties Division, Air Materiel Command at Wright Field: Should Parsons elect to continue patent prosecution, an assignment would have to be made to the government of the entire rights to the invention, subject to the reservation of a nonexclusive and royalty-free license to Parsons (3–2).

The matter of the Parsons Corporation releasing funds for completing preparation of the application was resolved near the end of 1951, more than nine months after the termination of the Parsons–MIT Agreement. At a Servo Lab meeting on November 5, John Parsons stated that he had obtained funds from his company for prosecution of the patent and was very much interested in moving forward. Although this meeting preceded Attorney Melvin Jenney's November 8 telephone conversation with Parsons' patent attorney, word about the availability of funds evidently had not reached the latter attorney at the time of Jenney's call.

All was now clear for filing the patent application. However, a little more than ten years was to elapse before the patent was issued, there being many communications with the patent examiner, the filing of several amendments, and many additions and deletions of claims in the intervening time. Because the patent was thought to be a major factor in future determinations of numerical control rights, *Control Engineering* magazine, in its April, 1963 edition, published an historical summary of the application's progression through the Patent Office, as well as some information on patent processing in general. The reader is referred to this publication for further details (3–3).

Forrester et al. assigned their patent through MIT as intermediary (a *mesne* assignment) to John T. Parsons, who in turn sold his rights to the

Bendix Corporation. Bendix then proceeded to execute royalty agreements with machine tool builders that were offering numerically controlled machine tools in their product lines. We recall many visits made by patent attorneys representing diverse organizations who wanted to search through our files for clues that might form a basis for setting aside the patent. Evidently none found a basis for proceeding, and the patent remained intact.

The assignment of the Forrester patent to John T. Parsons, rather than to the Parsons Corporation as had been specified in the original Parsons–MIT Agreement, caused later concern when patent attorneys for Giddings and Lewis delved into G&L's rights under its Agreement with MIT. Parsons, however, had worked out an arrangement with his company at about the same time that the Forrester application was filed, in which rights to the invention would be turned over to him. In turn, Parsons agreed to pay 12 percent of any royalties above $20,000 to the inventors as a group.

While the application for the MIT invention was being prepared, John Parsons was proceeding independently with an application for a patent on his idea for a punched-card controlled machine tool. This application, entitled *Motor Controlled Apparatus for Positioning Machine Tool*, was filed on May 5, 1952, three months before the MIT filing. The inventors were John T. Parsons and Frank L. Stulen, and the patent was No. 2,821,187, issued January 14, 1958. The Parsons–Stulen invention differed in many respects from the MIT invention. Machining information was stored on punched cards, the controller was mechanical rather than electronic, and because the machine could cut along only one principal axis at a time, it included a tiltable and rotatable platen, described as a "two-axis sine plate," which enabled straight-line cuts to be made at angles with respect to the principal axes of the workpiece. To the best of our knowledge, no machine tools were marketed under the Parsons–Stulen configuration.

4

THE MIDDLE YEARS

The middle years of the project, 1953 to 1959, were years of transition for our staff and years of progress for numerical control. After a slow start, the concept gradually took root in industry and was well along as an established product line at the end of the decade. A major effort beyond proof-of-feasibility was required by Servo Lab, the Air Force and industry, however, before that goal could be fully realized.

During the mid-1950s, the nature of the project changed markedly. Upon successful demonstration of the experimental machine, the staff turned its attention to an intensive information-dissemination program to familiarize industry with the potential of the new technology. At the same time, the development of another digital director system was undertaken at the request of the Giddings and Lewis Machine Tool Company. An economic study, which had been authorized by the Air Force in July, 1952, was also pursued in order to gather factual data on NC machining costs. The composition of the staff changed dramatically, too, as the project effort turned away from equipment development and concentrated on programming for NC. Software engineers replaced hardware engineers. These events are discussed in this chapter.

The Information Dissemination Program

After the Laboratory demonstrated its experimental NC machine tool in 1952, the Air Force hoped that, having sponsored the development through the proof-of-feasibility stage, it could withdraw and leave further developments to the machine-tool builders. In fact, during the demonstration conference, Colonel Wilbur R. Carter, who was then head of the Manufacturing Methods Branch of the Air Materiel Command, openly urged machine-tool builders to pick up the development and turn it into marketable products. The machine-tool builders were not ready to do that, however. Their reactions in the 1952–53 time period ranged from guarded optimism to outright negativism. Nevertheless, there were some isolated visionaries who foresaw the enormous potential in a machining system that could operate automatically, without manual intervention. A few were intrigued by the possibility of storing all the information needed to sculpture mechanical parts in a highly compact numerical form and by the ability to replicate parts over extended periods to the same degree of accuracy. Still others were attracted by the possibility of controlling machine tools remotely by electrical signals fed from a central digital computer to the machine tool over communication channels. However, the vast majority expressed skepticism, on a variety of grounds, about the practicability of the whole concept. Some questioned the reliability, accuracy, and economic viability of such a system, or were concerned that their organizations would not have staff with qualifications to maintain the equipment; nearly all were afraid of any system that used an abundance of vacuum tubes.

Fortunately, the Servo Lab staff and the Air Force had anticipated these concerns and the need to dispel them. Thus, two tasks, included in a supplement to the Lab's contract with the Air Force, negotiated in July, 1952, shortly before the demonstration, called for a program to familiarize industry with NC and for an economic study of the process. Actually, these tasks were intertwined. In order to gather reliable economic data on the machining of representative aircraft parts, the Laboratory solicited outside jobs, mostly from the aircraft industry. As part of the solicitation, there were, inevitably, discussions and presentations at the various organizations contacted.

Initial responses were discouraging. We were able to get representative aircraft parts for our economic study, but there were strong reservations about the process itself. To paraphrase McDonough, our staff members were always received graciously; they made their presentation; they were invited to lunch; they got to meet the president of the company; they left; and that was the end of it. Nevertheless, the visits left a telling mark, and attitudes gradually changed as the presentations helped industry appreciate the potential of NC.

The information-dissemination program also included an "open door" atmosphere at the Laboratory for persons from all over the world who wished to discuss the equipment and see it in action. Laboratory staff members made numerous presentations at professional conferences, published many reports and papers, and offered two one-week courses on numerical control, spaced a year apart, for engineers in the aircraft and machine-tool industries. The underlying goal was *information dissemination.* Throughout the indoctrination program, special attention was given to the aircraft industry because of the Air Force's mission to move metals processing technology forward as it applied to military aircraft manufacturing. As a university-based group, the Laboratory was ideally suited to carry out a program of this kind. It was unhampered by proprietary constraints and its information dissemination skills stemmed naturally from the educational environment in which it was immersed.

During the period 1953–54, visits to the Laboratory were very popular. A random sampling of our monthly progress reports to the Air Force indicates that about sixty persons saw a demonstration of the machine each month. This number occasionally doubled when a professional-society conference was being held in town. Nearly every month over the same time span, staff members visited different parts of the country to make presentations, the format of which was typically the showing of a motion picture film of the machine in operation, a presentation of machining results, and a general discussion with plant managers and operators. The information-dissemination program is discussed in further detail in Chapter 6.

The Economic Study

The economic study undertaken by the Laboratory had two components: the machining of approximately one hundred different parts, in varying numbers per part, and a more formal, controlled analysis of production costs associated with ten in the batch of one hundred. As part of the formal, controlled economic study, careful records were kept of times and costs incurred in the various stages of manufacture. To put realism into the program, the companies supplying the parts were billed for the jobs.

The basis for the study was this: although the early milling we did on our experimental machine provided tangible evidence of technical feasibility, it was concluded that carefully derived cost comparisons between conventional and NC machining would be necessary in order to establish economic viability in the minds of the users and machine-tool suppliers. Accordingly, during the summer of 1954, the Laboratory engaged Robert H. Gregory, Assistant Professor of Industrial Management, MIT, and Thomas V. Atwater, Jr, Vice President and Treasurer, Product Development Corporation, Boston, to undertake a detailed cost evaluation of the NC process in comparison to traditional machining processes. The salient features of the Gregory–Atwater study are presented here. Readers wishing more details should consult references 4–1 and 4–2.

The initial plan was to compare the cost of producing parts on the MIT milling machine with costs of the same parts produced by traditional methods. Ten parts were selected for detailed study from the hundred or so which Servo Lab had obtained from industry. These ten parts would be produced in the Laboratory, in various quantities, and cost figures would be sought from the companies that supplied the parts. It soon became evident, however, that although it was easy to get parts from industry, getting reliable cost data on them was impossible. Several reasons can be cited for this difficulty: many parts were experimental and had never been manufactured in quantity by any method; some companies were reluctant to release their cost figures, or their company accounting practices did not include costs for individual parts; or most of the parts were delivered to us by personnel who were too remote from the production practice to be able to estimate production costs accurately.

Hence, it was finally decided to obtain *estimates* of production costs for the parts from independent machine shops. Estimates of labor and machining costs for each of the ten pieces in the sample would be obtained from one or more of ten machine shops, and each shop would be offered payment at the going rate for its services. Seven of the ten accepted payments. The shops worked from blueprints and photographs of the pieces involved. A stipulation was that estimates would be made on the basis of making the part in lots of one to five and 25 to 30 pieces. All were told that participation with us would not lead to further business, either with us or with the organization that provided the part.

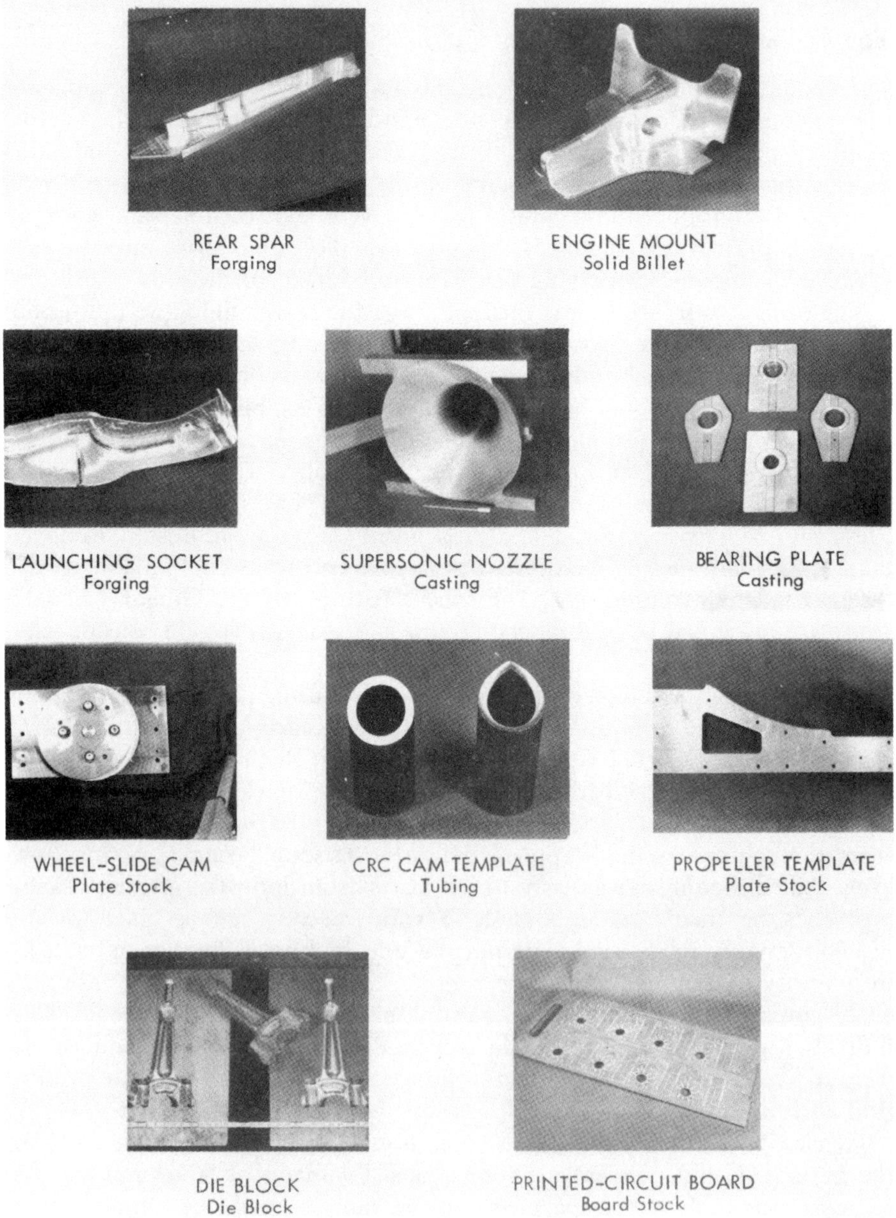

FIG. 4.1. The ten sample parts.

The comparative cost analysis examined both make-ready and operating costs. Make-ready tasks included planning for the method of manufacture, fixture design and manufacture, and, in the case of machining by NC, calculations required for determining tool-center loci, tape code preparation, and tape punching, checking and debugging. Operating costs included labor and related costs for setups, metal cutting, and hand-finishing (when needed), and an allocation cost to reflect depreciation in machine tool value.

There were some unresolvable variables in the study which may have had an influence on the results, but these were deemed insufficient to alter the general conclusions that were drawn. As examples, all parts machined by Servo Lab were actually made, and all were cut on its experimental NC milling machine—there was no opportunity to choose among machine types. In contrast, the outside machine shops provided their best estimates of costs, but no parts were actually made. The outsiders were free to designate whatever machine tool they would use to do the job.

Figure 4.1 illustrates the ten jobs chosen for study. The rear spar and the engine mount involved profiling and hog-out work; the launching socket, supersonic nozzle and a bearing plate cast in aluminum required finished machining from castings and forgings; a third category consisted of a cam and templates; and the fourth category, labeled miscellaneous, comprised a die block and a printed circuit board.

A general statement about results is that Servo Lab's per-part machining costs were lower than the outside estimates but Servo Lab's make-ready costs exceeded those of outsiders, principally because of programming and tape preparation costs. Since make-ready cost is a one-time occurrence for a part, it follows that NC would yield lower overall per-part costs whenever a sufficient number of parts was made.

This general statement requires further elaboration, however. Consider, for example, the rear-spar fitting. Informal estimates obtained from two shops showed overall costs below Servo Lab's, no matter how many parts were made. On the other hand, a careful, detailed, step-by-step estimate from a third shop showed that, although Servo Lab's make-ready costs were high, its production costs were sufficiently lower to yield a smaller total cost for as few as 12 pieces. Again, in the case of the propeller template, for which programming was relatively simple, Servo Lab's make-ready and per-part machining costs were both lower than the comparative figures supplied by the outside shop. On the other hand, the 5"×8"×8" engine-mount fitting was a complicated job of sculpturing from a forged billet of aluminum alloy; it required a large amount of programming time, and the cost of fixture design and manufacture about equalled that of programming. Servo Lab's overall cost showed up poorly for this part, even though its per-part machining cost was comparable to the outside estimate. In retrospect it is remarkable that this new technology, only three years after its inception and still untested in commercial production settings, was able to compete on a cost basis with time-tested machining methods that had evolved from decades of development and refinement.

By 1958 data on the economics of NC based on production floor experiences were becoming available. These data confirmed our results. For example, William D. Beeby of the Witchita division of the Boeing Airplane Company reported that: "Numerical control has proved it can reduce costs, reduce lead time, improve quality, reduce tooling, and increase productivity," and he backed his claim with shop-floor production data (4–3). Since numerically controlled skin mills were not available for production of early versions of the B-52G aircraft, template controlled machines were used in the first production runs of the sculptured-type skins of these aircraft. Authentic cost and time data for template control were thus available for comparison with later data derived from machining similar parts through use of NC. Beeby's figures were impressive. The amount spent on startup costs—planning, tooling, and other costs—for one sculptured-type skin was $18,467 for template control versus $1957.50 for NC. The time to machine the first part by template control was approximately 1400 man-hours and 750 man-hours for the ninth unit. In contrast, with NC the first part consumed 700 man-hours and the fourth unit, 300 man-hours. Equally impressive were Beeby's figures on tooling for a forging. For template control 42 fixtures were required at a cost of $37,772; for NC the number dropped to 14 at a cost of $7625. Additional comparisons are included in his paper. Noteworthy, though, is that Beeby provided no information on programming costs.

The Gregory–Atwater study brought into sharp focus the true role of continuous-path NC machine tools as a manufacturing method. NC showed excellent promise for producing parts for which demand ranged from a few to a few hundred. Thus a clear distinction was drawn between numerically controlled stand-alone general purpose NC machines and "Detroit automation," where concatenations of specially designed machines and automatic materials handling equipment would be economically justifiable because of the huge quantities of identical items that would be produced.

Gregory and Atwater also reinforced the conviction of the Servo Lab staff that even further economic viability would result if means could be found to reduce programming costs. Before addressing the Laboratory's programming research effort, we shall turn our attention to another piece of equipment development, the Giddings and Lewis project.

The Giddings and Lewis Project

The information-dissemination program had a salutary effect on the future of numerical control. On-site visits around the US provided an opportunity to clear up many misconceptions about the system that had arisen from word-of-mouth and news media information.

It became evident that it was possible to arouse interest by face-to-face contacts and open discussions. Still, after almost a year of "missionary work," McDonough wrote in his August, 1953 report to the Air Force: "It would appear from our present data that the electronic industry is much more interested in producing control equipment for machine tools than the machine tool industry is in producing special machines to be controlled." Nevertheless, noticeable progress was being made. In his October, 1953 progress report, McDonough, commenting on his visits to five aircraft companies in the Los Angeles area, was able to write: "Much of the natural resistance to change felt on previous trips to the West Coast has disappeared. There is still some reservation about the electronics involved in numerical control, but most of these companies appear willing to accept the electronics if the advantages of numerical control are forthcoming too."

Although more skepticism was encountered among machine-tool builders than from the user group (the aircraft companies), one machine-tool builder, the Giddings and Lewis Machine Tool Company (G&L), located in Fond du Lac, Wisconsin, was a notable exception. Giddings and Lewis had followed MIT's numerical control project closely, beginning with the MIT demonstration in 1952, but as Edgar McFerren, a G&L vice president, expressed it: "We didn't really catch fire on it [the concept] until we received an order from Wyman–Gordon for a big horizontal boring mill" (4–4). This order was placed under Wyman–Gordon's contract with the Air Force for a 75,000-ton aluminum press for making large structural aircraft components. The mill G&L proposed to build was, in essence, two machines on one runway, which would machine the dies for the Wyman–Gordon press using tracer-type duplicating equipment. These were huge dies—40 feet long and 10 feet wide—which would be set up on their sides during the machining of the cavities. In this configuration, G&L was confronted with the fact that the stylus for following the master had to be 12 feet or so above the cutting-tool spindle (way up in the "breeze," as McFerren explained), with the resultant possibility of large machining errors because of structural flexures.

In retrospect, we see here an example of how a new demand on a conventional technology can push the technology to its ultimate limit of capability —in this case, probably beyond its capability. At this point it often becomes necessary to turn to an entirely new way of thinking and to start along an entirely new track. In any event, the Air Force's heavy-press program was in trouble because the mechanism for precision-machining the press dies was in trouble. Fortunately G&L turned its thinking toward numerical control as a way to get within the 10- to 15-thousandths of an inch tolerance they were seeking.

McFerren also had another reason for becoming interested in NC: he sensed a movement in the aircraft industry toward numerical control as G&L representatives met with users and other builders to discuss specifications for the machine tools to be purchased under the government's upcoming long lead-time procurement program, described in detail in Chapter 6.

As part of Servo Lab's information dissemination program McDonough and Richard Lawrie visited G&L at Fond du Lac on January 30, 1953. In his report on this meeting, Lawrie indicated that it was initially apparent that G&L representatives at the demonstration felt NC had no immediate advantage over existing controls because of the complexity of the director.

However, as the meeting progressed, G&L's fears and reservations disappeared and enthusiasm gradually developed. On March 10, Lawrie received a phone call from McFerren inviting the Laboratory staff to Fond du Lac for a second presentation and to discuss a possible study contract with MIT. This meeting, held on March 19–20, 1953, culminated in an informal agreement to proceed as quickly as possible with a study of the applicability and adaptability of NC to the analog magnetic playback system being supplied to G&L by the General Electric Company. A study cost estimate of "between $30,000 and $40,000" was discussed and agreed upon, and a formal Memorandum of Agreement was signed on April 3, 1953 (MIT's contract DIC 7108).

The first report to G&L was verbal (a final report was never written) and indicated the application they had in mind was indeed compatible with NC. There was sufficient evidence at that point to convince G&L to proceed with the building of a digital control system under a new contract with MIT (DIC 7163) signed August 25, 1953, and formally announced at a meeting with General Electric engineers at Schenectady, New York on September 2, 1953.

Although G&L was initially attracted to NC through its Wyman-Gordon

FIG. 4.2. The G&L 5-axis skin milling machine and its associated magnetic-tape unit. Courtesy of the Giddings and Lewis Machine Tool Company.

contract, the exact machine for which Servo Lab's G&L director system would apply was left open. There was, however, a requirement that the director should be compatible with a magnetic-tape playback control system G&L had under development at the General Electric Company. In other words, the concept was this: a G&L machine tool would be driven from electronic signals stored on magnetic tape and the stored signals would be either tracer-generated or derived from MIT's digital director. Either way, the signals driving the machine tool would be identical.

As it turned out, the MIT director and the GE playback unit were demonstrated on a G&L bridge-type, 5-axis skin milling machine. Since the director was *off line*, meaning that its output signals were stored on magnetic tape rather than applied directly to the machine-tool servos in real time, one director could serve any number of machine tools and their associated playback units. The five-axis machine tool and its magnetic-tape drive are shown in Fig. 4.2. The machine had a 6'×18' movable table and two head stocks, each with motion along two axes, giving five axes of motion in all. The director system built by Servo Lab is illustrated in Fig. 4.3, together with the specially designed paper tape key-punching desk. Observe that the input to the director is paper tape; output signals are recorded on the mag-

FIG. 4.3. The G&L director unit and its paper tape preparation desk. Courtesy of the Giddings and Lewis Machine Tool Company.

netic tape unit shown at the right of Fig. 4.3. The system was advertised by G&L as their Numericord system. Although G&L looked upon the machine-tool system illustrated in Fig. 4.2 as experimental, it was capable of machining actual aircraft components, and G&L used it to make a small quantity of parts for the tail section of North American's F-100 aircraft.

The original Memorandum of Agreement signed with G&L for the study project was broadly stated and called for "the development of control-system specifications for prototype machine tools through an extension of the techniques of numerical control." Professor William M. Pease was designated as the research supervisor. As previously noted, the study gradually blended into a follow-on contract for designing and constructing the director system. The new contract designated Professor J. F. Reintjes as supervisor, Pease having terminated his position at the Institute in the interim. Contract amounts were left pretty much open-ended. The study Agreement called for an expenditure of $30,000, but design work began before these funds were depleted. However, the original oral estimate given to G&L of $150,000 for fabrication of the director was grossly understated for several reasons: new design features were built into the system; as the work progressed, every effort was made to make the unit a final prototype which could be duplicated and marketed; on-site setup of the director at Fond du Lac was not anticipated at the outset; and no thought was given to the need, which developed later, for reproducible engineering drawings. The total amount billed to G&L for the entire project was $390,056.63.

As the unit neared completion, there was enormous pressure on our staff to ship it to Fond du Lac and get it working there with the five-axis skin milling machine and playback unit. G&L had firm plans for a public demonstration in June, 1955 and for a few refinements before the September Machine Tool Builders' Show. The Laboratory was able to deliver the director at Fond du Lac toward the end of May, but our staff had difficulty getting it to run reliably, mostly because of thermal expansion problems with the feed-through rivets in the printed circuit boards. Between the heating and cooling cycles, the soldered joints at the rivets would break loose and disrupt continuity through the boards. A near-calamitous fire also broke out in the magnetic-core storage units during debugging. Evidently the enamel insulation on the wire used for the windings contained nicks; eventually the short circuits that developed at the nicks caused the cores to burn up. A graphic description of the event was provided by Susskind, who was in charge of the electronics. He said: "I'd come from Cambridge to get this thing ready for the demonstration. I walked into the room, and there's a whole gang sitting there like a bunch of old ladies with sewing needles rewinding the cores." Nevertheless, "the gang" got the whole system working—at 3:00 AM the night before the 9:00 AM demonstration which was held over a four-day period, June 7–10, 1955, at the Fond du Lac plant. The invited guests included 118 representatives of the aircraft and electronics industries, the military, the technical and popular press, and the academic community. Although the system was not exhibited at the Machine Tool Builders' Show held in Chicago the following September, G&L arranged to fly interested parties to Fond du Lac for on-site demonstrations.

The G&L director was substantially different from the Laboratory's original director, partly because of different specifications and partly because of advances in component technology. As indicated above, the director ran off line and its output signals were stored on magnetic tape. Instrument servos were no longer used, and the digital coding was in binary-coded decimal rather than straight binary, a feature that simplified tape preparation. The G&L director substituted magnetic-core storage for relay storage; in fact, relays were used very sparingly, since experience had shown they were fre-

FIG. 4.4. The printed circuit board construction used in the G&L director. Courtesy of the Giddings and Lewis Machine Tool Company.

quently the source of trouble in the original director—more so than the vacuum tubes. Some additional system design changes were also necessitated because of the requirement that the signals supplied by the director be compatible with those contained on GE's tape. As indicated in Fig. 4.4, rather than using breadboard, experimental-type construction, the director was built as a professionally packaged unit; most of its electronic circuits were mounted on removable printed circuit boards with rear plug-in connectors, and elaborate precautions were taken against overheating through installation of a filtered forced-air ventilating system. As noted previously, the tape preparation equipment was specially designed to handle the key-punching and encoding operations and was incorporated into a customized tape-preparation desk.

Patents

The patent agreement between MIT and Giddings and Lewis was more complex than those between the Institute and Parsons and between the Institute and the Air Force. In the patent clause, it was necessary to acknowledge the existence of the other two agreements. This was accomplished by giving recognition to the "indivisible relationship" between the G&L contract and "the extensive Government contracts undertaken by MIT in this field." As a result, G&L agreed to grant to the government a "free, irrevocable, non-exclusive license for Government purposes to practice any invention reduced to practice in the course of the [G&L] contract." A further stipulation was that certain unpatented but possibly patentable ideas that had already evolved under the MIT–government contract would belong to MIT and the government, but G&L would have a royalty-free non-exclusive license to use any such patentable idea that would be adapted to the G&L director.

Although the MIT–G&L agreement clearly addressed the patent situation between the Institute and the government, it was silent on MIT's patent agreement with Parsons, a matter that caused some concern when G&L's patent attorneys began to delve into the situation toward the end of 1953. It is even fair to say that at a January 12, 1954 meeting at MIT, G&L was openly perturbed that the Forrester patent, based on the experimental machine, was pending, and that, as a result of an inquiry directed to Parsons' attorney, G&L had discovered that Parsons was negotiating with Bendix and others to sell an exclusive license, both within and outside the machine tool field—with a price tag of half a million dollars. Since no licensee had yet been chosen at the time of G&L's inquiry, Parsons was willing to talk with G&L too.* The question was also raised at that meeting whether information on the status of the Forrester–Parsons patent application had been withheld from Giddings and Lewis, but this notion was dispelled when all agreed that Pease, in his initial dealings with G&L, had fully apprised G&L's president of Parsons' patent rights. Evidently the implications of those rights as set forth by Pease were subsequently set aside or forgotten.

*G&L was not successful in its attempt to gain an exclusive license from Parsons.

The numerically controlled machine tool system developed by Servo Lab for G&L represented such a significant advancement over the original Servo Lab system that it was decided to file a patent application for it even though other NC applications were pending. The inventors of the G&L system were James O. McDonough, Alfred K. Susskind and Herbert P. Grossimon, all Servo Lab staff members. The patent (No. 3,390,315), issued June 25, 1968, was entitled: *Apparatus for Numerical Control of a Multiaxis Machine Tool including Interpolation and Feedrate Control*, with a total of 146 claims allowed. The basis of the invention was the many improvements over the original experimental NC machine tool which were incorporated into the G&L director system. Rather belatedly, G&L succeeded in licensing the General Electric Company under this patent.

A second patent was also issued under the MIT–G&L project. It pertained to controls and associated equipment used in preparing paper tape for the G&L director, and was issued December 6, 1960, as No. 2,963,137, entitled: *Controls for a Typewriter and Associated Apparatus.* The inventors were James O. McDonough, Alfred K. Susskind, Herbert P. Grossimon, and Francis F. Lee—again, all Servo Lab staff members who had worked on the G&L project.

Programming

We now turn away from the G&L project and go backward in time to take up the subject of programming research. Our purpose is to trace the evolution of programming techniques for NC machines from the early days of manual programming to the adoption of the automatically programmed tool (APT) system as a world standard. Even before the Laboratory's experimental machine tool became operational, it was evident that programming

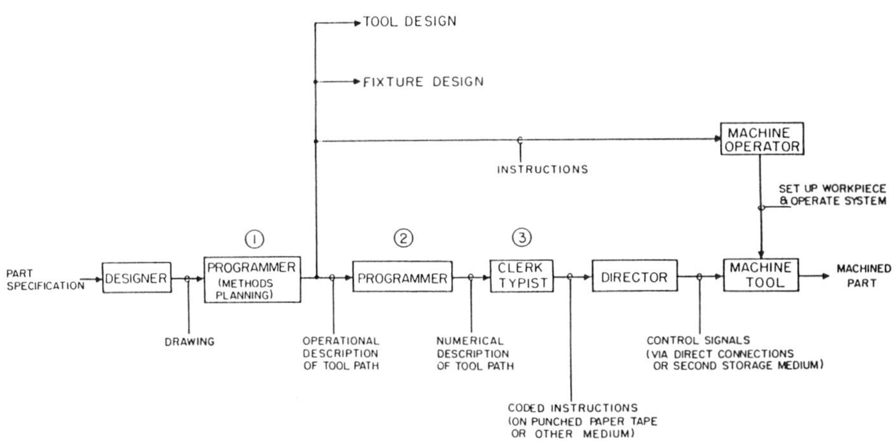

FIG. 4.5. Information processing for a manually programmed numerically controlled machine tool.

the machine would be time consuming, and that programming time was highly dependent on the complexity of the parts to be machined. This conclusion was borne out when the time came to program the parts that had been solicited from industry. For most of the ten parts in the Gregory–Atwater study, the sum of programming and tape preparation times exceeded machining times by factors that ranged from 18 (the die block) to 52 (the propeller blade template). A notable exception was the bearing plate, which was easy to program because only a few long, straight-line cuts were involved. For this part, machining time equalled programming and tape preparation time, about two hours each; of the two hours of machining time, only seven minutes were required to complete the cutting. The rest of the time was consumed in setting up the hold-down fixture and the workpiece.

In order to put programming into perspective, consider the steps required to produce a part by numerically controlled means. The process is illustrated in Fig. 4.5. It is assumed in this chain of events that programming is done manually. Observe in the figure that the programmer, as well as the machinist and the tool designer, enters the process at the methods-planning stage (indicated by the encircled 1). At this point, decisions are based on information derived from a drawing (or drawings) of the part to be machined. The decisions include the number, order and directions of the cuts to be made, the types and sizes of the cutting tools that will be used, the design and mounting requirements of the hold-down fixtures, tool spindle speeds, and initial setup instructions.

The second step in the process (the encircled 2) is the programming stage. At this point, the programmer takes the verbal descriptions of the tool cutting-paths generated in step one and translates them into numbers. This translation is heavily dependent upon the design configuration of the digital director, particularly with respect to the way the director handles curved surfaces. In the MIT experimental director, straight-line (linear) interpolation of curves was used, but other techniques could be employed, for example polynomial interpolation.* With linear interpolation, the process of developing numerical descriptions of a cut consists in finding the cutter locations that must be specified in order to maintain error tolerance, computing them, and tabulating the values together with the corresponding feed-rate data. This stage of data processing requires mathematical calculations—especially of plane- and analytic-geometric relationships—that provide straight-line approximations to curvilinear surfaces. In manual programming, these calculations were performed with the aid of an electromechanical calculator. The calculations for planar (two-dimensional) cuts turn out to be simpler than those for three-dimensional surface cuts, although both are influenced by the kind of cutter being specified: ball-end, cylindrical, or other type. Bear in mind also that the numbers being specified are those for the *cutter center*, which must be offset from the desired surface by the cutter radius.

---

*For a detailed discussion of interpolation methodologies and the mathematics involved, see reference 4–5.

Another part of the programming process at stage two is the specification, for each straight-line cut, of the distance of the cut along each axis and the cut time. Should slowdown be required near the end of the cut in order to avoid overshoot, the total cut is divided into two or more segments, and new distances and times are specified for each segment.

The third step in the programming process (the encircled 3) consists in translating the numerical description of the machine's motions into the particular coded form being employed in the digital director and recording the corresponding data on a suitable storage medium—paper tape, in the case of the first experimental machine and the G&L director. The data per cut to be key-punched for these machines would be the difference between successive tool-center coordinates and the time required between the beginning and end points of the cut. It was found that a clerk typist could handle this task. Thus the programming operation ends with a complete "road map" description, in digitally encoded form, of cutting-tool motions, stored as a pattern of holes punched onto a paper tape. To complete the process, the tape is fed to the digital director which drives the machine tool.

From the preceding discussion, it should be evident that a general purpose digital computer should be considered as a means for carrying out steps two and three of the programming process. Since these steps are largely mathematical, they should be ideally suited for machine computation, with a resultant saving in programming time and cost over that incurred in manual programming. The Servo Lab staff recognized very early the bottleneck posed by programming and began to investigate the potential of computer-aided programming. In fact, a study of machine aids in paper tape preparation was initiated during June, 1951, nine months before the experimental machine tool system was put into operation. A graduate student, John H. Runyon, undertook an investigation of how MIT's Whirlwind I computer might be

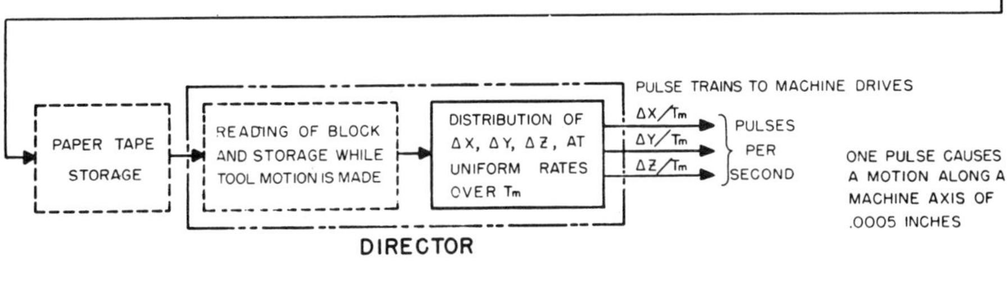

FIG. 4.6. The role of a general purpose computer in programming the experimental numerically controlled milling machine.

used as a programming aid. His research, the results of which were used in the thesis he submitted for the Electrical Engineer degree (4–6), was an important early milestone in the evolution of automatic programming. As indicated in the top set of boxes in Fig. 4.6, Runyon proposed that a general purpose computer be used to carry out many of the mathematical operations formerly done manually with the aid of an electromechanical calculator, and he backed up his thesis by writing a substantial number of general programs, called *library subroutines*, that could be drawn upon from job to job, as the occasion required. His library included subroutines for computing tool-center offsets, determining cut length and cut spacing, providing feed-rate control and a computation of minimum clock cycle, and encoding instructions for two specific types of surfaces, namely airfoils and cones. To utilize these subroutines, there was also required a library of subroutines that adapted the Whirlwind computer to the specialized requirements of the experimental milling machine director. All subroutines were stored on punched paper tape (Whirlwind's input program storage medium), with each subroutine identified by a number that allowed it to be manually selected from the library as needed.

The significance of Runyon's work was that it revealed the power of the digital computer as a programming tool for numerical control and demonstrated how development of a library of subroutines could further enhance the power of the tool by eliminating a high percentage of the repetitive coding required for each new job. Runyon was able to show that his subroutines reduced the amount of *new* computer code by amounts ranging from 75 to 90 percent.

Although Runyon's work represented a major forward step in automatic programming, it had limitations. It worked best for two-dimensional cuts and three-dimensional surfaces that could be defined by relatively simple mathematical expressions. To handle a warped surface of arbitrary shape, however, subroutines would be needed that would likely have little in common with other warped surfaces. Hence, for such shapes, the idea of storing a library of subroutines that could be drawn upon for a variety of machining jobs would lose its luster.

We also see in the library subroutine concept an approach to automation that very often falls short of what is truly required to put an automated process in a superior position. The library subroutine concept, in essence, automated the repetitive steps and calculations of manual programming: it preserved the steps of manual programming, but dealt with certain parts of it through use of a general purpose computer run by sets of routines that were in turn synthesized from an assortment of previously prepared subroutines. The utility of this approach was fine so long as the subroutines were applicable to an abundance of different parts but their value vanished for surfaces that would unlikely recur either within the same or different parts. Clearly the whole programming process needed to be reexamined from a system viewpoint in light of the use of the digital computer as a programming aid.

A step in this direction was underway even before Runyon had completed his investigations of the library subroutine approach. The focus of this advanced work was in Forrester's Digital Computer Laboratory, a spin-off of Servomechanisms Laboratory, but still closely coupled to it. When facilities required to house the Whirlwind I computer exceeded those which Servo Lab could provide, Brown and Forrester agreed to set it up in an off-campus building and to designate the new facility as the Digital Computer Laboratory. A staff member of that lab, Arnold Siegel, became intrigued by the idea of automating the NC programming process along lines different from the library subroutine process, and he undertook a study of automatic programming for NC under sponsorship of Servo Lab's NC project.

From Siegel's work there emerged a new concept in programming for NC. Siegel sought to simplify and automate the approach in two ways (4.7, 4.8): He suggested transforming blueprint information into machining commands stored on paper tape by first giving input instructions to a general purpose computer in a language that consisted partly of English words and partly of alphanumeric expressions, and then programming the Whirlwind computer so that it recognized the instructions, executed them, and prepared as output a complete paper tape that would drive the digital director of the experimental NC machine tool. A few examples will serve to illustrate the form of the input command language developed by Siegel. To indicate the use of a cutter of a certain diameter, say ¾", and to tell the machine that the cutter should move to the right of the work, Siegel would issue the commands:

```
TOOL RADIUS=0.375
RIGHT.
```

Machining tolerance was specified by the TOLERANCE command, followed by the numerical value of the tolerance, thus:

```
TOLERANCE=0.0005.
```

Points, lines and circles were defined by *tags*, followed by an appropriate number of integers. Thus, tags *p*, s, and *c* signified *p*oint, *s*traight line, and *c*ircle, respectively. The integer associated with each symbol designated destination points on the workpiece with respect to a reference point. Cutter speed was designated simply by supplying the appropriate number.

These various cryptic words, symbols and numbers could be combined as the program required. Thus, in Siegel's input language:

7.5, p3	meant that the cutter should move at a rate of 7.5 inches per minute in a straight line to point 3 on the workpiece; and
15, +c2, p2	would indicate the cutter should move at 15 inches per minute along a circle, c2, in a positive (+) direction, to point p2.

The second part of Siegel's work consisted of a set of programs that could accept his input commands, automatically carry out the repetitive math-

John E. Ward

ematical operations involved, and express the results on a perforated paper tape in a form recognizable by the machine tool's digital director.

Siegel's contribution was clearly another forward step in automatic programming. It made the human programmer's job much easier to learn, and greatly reduced not only the programmer's time but also the overall time required to prepare a complete paper tape. For example, Siegel reported that one mechanical part requiring about eight hours to program by hand was programmed for further processing on the Whirlwind I in less than 15 minutes. Whirlwind's time to do the calculations and produce a tape was 47 seconds. Still, Siegel's system also had limitations: it was confined to two-dimensional surface (planar) machining and it was close-ended, meaning that it was not readily expandable to three-dimensional surfaces. Hence, still more research was needed.

The Automatically Programmed Tool Project

The ultimate solution to automatic programming came during the years 1956–59, in the form of the automatically programmed tool (APT) system, the brainchild of Douglas T. Ross. A mathematics major at MIT, Ross came to Servo Lab in 1952 to work on several projects in the realm of human–computer systems for high-speed data processing, one of which was the digital flight test instrumentation (DFTI) project, headed by John E. Ward. The concept of DFTI, an Air Force sponsored project, was that performance measurements of avionics equipment would be made during inflight testing and stored on-board in digital form. Flight-test results would be compiled by means of a digital computer on the ground immediately upon

Douglas T. Ross

completion of a test. Ross's job for DFTI was to develop the software required to reduce, on a UNIVAC 1103 computer, the digitized measurements made during in-flight testing of the tail-defense system for the B-58 aircraft. The DFTI project was coming to a close at about the same time there was a need for a take-charge, computer software type person with fresh ideas on automatic programming for NC. Fortunately, Ross was available to take on the job, and he proved to be the right man in the right place at the right time. Ross had a very fertile, creative mind and a flair for independent thinking; he possessed the ingenuity needed to solve a complex programming problem and a keen sense of what was required to put a new concept across. From Ross's work emerged APT: the automatically programmed tool system for programming numerically controlled machine tools.

The development of APT was a major turning point in the evolution of NC, for it settled once and for all the issue of whether or not NC could be made economically viable in light of programming costs. Still, the project's future was held in a delicate balance, even before it got underway, as John Ward explained in a letter to Ross several years later:

> One thought occurred to me [in reading your APT language history]—and that is, how close the Lab came to not taking on the project. I remember sitting in your office with you and Frank [Reintjes], trying to decide what to do with the contract received from Wright Field on the basis of a proposal submitted by McDonough [and his group], who were no longer there. The question to be decided was whether a team could be put together to do anything with it, or whether it should be returned. As I remember, the balance between these two options was fairly even, but we finally agreed to go ahead. . . . (4–9)

That was a memorable meeting because it required a critical, objective assessment of the proposed effort in light of the purposes of an academically oriented research laboratory. Since technical feasibility of NC had been demonstrated, and some progress in automatic programming for it had already been made, there would be ample justification for the Lab's withdrawal from NC and the diversion of its resources to new endeavors. Also, the fact that the group which had developed the hardware under McDonough's leadership had resigned from the Lab to form its own company, Concord Control, Inc., to manufacture digital directors, made 1956 appear to be a convenient stopping point. Hence, the tests which would have to be met in order to merit further research on automatic programming along new lines were these: was the research sufficiently challenging and important to warrant our undertaking it? Was industry able to, and interested in, doing automatic programming research on its own? Was a person available to lead the project? Would there be opportunities for student participation? Was funding available?

As indicated in Ward's letter, funding was never a controlling element in the decision-making process: funds were in hand. The issue of research importance, and industry's readiness to undertake it, raised interesting questions, however. Siegel had made inroads on an approach to automatic programming, but we knew his results, while promising, fell short of what industry needed. We also had information that at least one aircraft company, the Boeing Company, was heading in the direction of library subroutines, a dead end. Hence, there was clear-cut evidence that the problem was far from solved and that it merited our attention. Since a highly creative person, Ross, was available to lead the work, we decided to structure the venture around him and to provide some graduate assistants to get him started. In the fall of 1956, two graduate students, Sam Matsa and George Zames, were assigned to the project as research assistants. They joined Harry E. Pople, Jr, and Jerome Wenker, who had started during the summer as full-time research staff members. Thus the NC project at the Laboratory headed in a wholly new direction, with a new complement of personnel.

One major hitch had yet to be resolved before we could make the final decision to proceed. Some of Ross's time was needed to wrap up loose ends on the DFTI project, and he wanted to be free to address other challenging applications of programming that might come to his attention. More specifically, Ross was eager to utilize his technical leadership qualities, but he had no enthusiasm for the administrative details associated with running a project. We settled the matter by setting up a Computer Applications Group in the Laboratory, designating Ross as its head, and establishing John Ward as the acting project engineer for the automatic programming project. Later another Servo Lab staff member, Donald Clements, replaced Ward as the permanent APT project administrator.

Ross recognized both the strengths and weaknesses of Siegel's approach to automatic NC programming. He concluded that the persons doing the part-programming should be able to express their machining instructions in a simple, English-like language, but he also believed the vocabulary would have to be expanded far beyond Siegel's short list of commands. More importantly, he felt the programming system should be broadly structured, so that it would be expandable and capable of growing with the experience gained through its use. In other words, it should be *open-ended.*

Ross's ideas for programming went far beyond programming for numerically controlled machine tools. He envisioned programming in general as an interrelational process between human and computer in which a general purpose computer would, in effect, be turned into a special purpose machine through use of a set of special programs developed to deal with whatever problem was at hand. The human would then work back and forth with the computer, in a conversation-like mode, in order to arrive at an overall program capable of handling a particular *class of problem.* The key to the success of this approach resided in the design of an appropriate language for the interchange of ideas. As Ross expressed it: "The language must bridge the gap between the fundamentally incompatible characteristics of the two parties. The human is quick-witted but slow, while the computer is slow-witted but extremely fast." (4–10) In a discussion between the author and Professor Fernando J. Corbato, a pioneer in the concept of timeshared interactive computers, Corbato paid tribute to Ross's contributions to the art and science of programming by indicating that it was Ross who really redirected the thinking of system programmers away from the well-established machine language approach toward hierarchically structured computer languages in which major parts of the programming job are delegated to the machine itself.

Ross's APT system indeed turned out to be a general framework that made possible orderly additions of features which would improve the power of the system. Regardless of its evolutionary state, however, the goal was always the same: namely, to permit a person with no programming skill to give instructions for machine-tool motions in easy-to-learn English-like terms and to relegate to a general purpose digital computer the jobs of translating these instructions into a language that would be understood by the computer; making the mathematical and geometric calculations required to implement the instructions; and post-processing the numerical calculations and instructions in a way that would be understood by any one of several different digital directors. It should be noted that, at the time the APT development program got underway, machine tool builders had begun to design their own machine tool digital directors, not all of which had identical characteristics. Hence, to preserve the generality of APT, a flexible post-processor was essential.

Although the above goal is simply stated, achieving it in the generalized, open-ended fashion that Ross envisioned was a major job that required development of a wholly new computer language system. The new language had to be expandable, easy to learn and use, and internally consistent as it was expanded, with respect to vocabulary and to semantic and syntactic structure. It is beyond the scope of this treatise to present details of what emerged as the APT language; it has been amply described in the literature (see Bibliography: Reports, item 1). Suffice it to report here that APT was not a single, specific programming system; rather, it was a family of systems that evolved over time, with increasing degrees of sophistication in the way it described surfaces that were to be contoured by cutting tool motions and in the way it processed information. Ross described these programs as "three styles of three-dimensional programming": programming by points, programming by space curves, programming by regions. Surface solutions were expressed as normal vectors and directed distances rather than as outputs from a library of subroutines.

Although Ross and his group were not familiar with machine-tool technology and the technology of controlling machine tools by numerical methods when they took over the project, they lost no time in acquiring competence in these matters and relating them to what they thought the programming project should accomplish. Toward the end of 1956, approximately six months after project startup, a rudimentary structure for APT was in place and had been described in the first interim project report (4.11). It included such basic APT features as surface terminology, the concept of turning a general purpose computer into a special purpose machine through a special purpose programming system, and an outline of generalized structural requirements for arithmetic calculations. It is noteworthy that Ross came up with the acronym APT in mid-December of 1956, while he was drafting his first interim report (4–12).

It was clearly evident at the beginning of the APT project that available funds were unrealistically low in light of the magnitude of the programming effort that would be required. We were also aware that taking on an underfunded research project is risky, particularly when the research is of the go/no-go type, that is, if no partial demonstrable results can be produced until all the work has been completed. From a sponsor's viewpoint, an element of doubt exists about the wisdom of continuing such an effort, as periodic requests are made for additional funds, and hence there is a temptation to terminate the effort and cut losses. Fortunately, APT was structured as an evolutionary concept, with periodic milestones that could be used to show partial results and steady progress. Fortunately also, our Air Force project engineer, William Webster, had confidence in our staff and its ideas for automatic programming; he saw to it that both his superiors at Air Materiel Command and users of NC equipment in the aircraft industry were apprised of our project plans. Thus, during the summer of 1956, Webster brought his superior, Colonel Dick, to the Laboratory to hear firsthand how we proposed to tackle automatic programming. Four months later, in November, Webster alerted us that several aircraft companies, through the Aircraft Industries Association's Subcommittee on Numerical Control (SNC) were grappling with the NC-programming issue. Webster recommended that we contact the chairman of SNC, and also Kenneth Wood of Boeing, Seattle, who was involved in Boeing's plans for NC programming.

The upshot of these industry contacts was that Ross received an invitation to present his views on automatic programming at a four-day SNC meeting held in Dallas, January 21–25, 1957. Armed with the contents of his first interim report,which outlined his whole philosophy of automatic programming, and imbued with enthusiasm and full confidence that he could successfully implement a system reflective of his philosophy, Ross proceeded to establish Servo Lab and his Computer Applications Group as a focal point for resolving the crucial matter of programming for NC.* From the Dallas meeting there emerged a close relationship between Servo Lab and the AIA that would endure for the next three years. Indeed, this relationship proved vital to the success of NC, because AIA's Subcommittee on Numerical Control (SNC) eventually accepted APT as the way that the aircraft industry should go, and the Subcommittee also accepted our invitation to help write the programs. As has already been pointed out, the project was so under-funded that we would have been unable to complete an operational APT system in time to install it as the industry standard without outside help. Since the government's long lead-time machine-tool procurement program was well underway, recipients of the NC tools in this procurement had to decide quickly how the programming should proceed. The procurement program is described in detail in Chapter 6.

Although Ross made a strong impression on the SNC membership at its Dallas meeting, the APT concept was not immediately embraced to the total abandonment of all other approaches. The subcommittee itself had problems: SNC had to find some way to avoid chaos in the programming language domain while at the same time providing flexibility that would allow individual companies to give expression to their programming ideas. Therefore, SNC continued to receive, discuss and evaluate proposals from their SNC programming taskforce during the several months following the January meeting. Servo Lab contributed to this evaluation process through Ross's many communications with SNC members, his presence at their monthly meetings, and most importantly by organizing a special eight-day course at

---

*The Aircraft Industries Association (AIA), now known as the Aerospace Industries Association, has a 70-year history of cooperation among its members and with the government. The AIA can trace its origin to a predecessor organization, the Aircraft Manufacturers' Association (AMA), The AMA was founded in 1917 to deal with wartime aircraft production problems, especially those relating to cross-licensing of patents among aircraft companies, cross-licensing being deemed mandatory for attainment of maximum wartime production in this newly emerging industry. The organization, also known as the Manufacturers' Aircraft Association (MAA), was dissolved after World War I and replaced in 1919 by the Aeronautical Chamber of Commerce (ACCA). The ACCA had an ambitious charter; it's purpose was to "foster, advance, promulgate and promote" aeronautics and "generally do every act and thing which may be necessary and proper for the advancement" of American aviation. At the onset of World War II, it was the ACCA that helped to establish East and West Coast War Production Councils which coordinated industry support with the War Planning Board and the military services.

At the close of the war in 1945, the name was again changed, to the Aircraft Industries Association. As the industry's first full-fledged trade association, the AIA's activities included not only the promotion of the technical and commercial interests of the nation's aircraft builders and suppliers, but also a liaison role between the government and the industry on technical and contractual matters. Finally, in 1959, the name was once more changed to the Aerospace Industries Association in order to reflect the movement of the industry into the space age.

Officially, the AIA of America describes itself as "the national trade association of companies in the United States engaged in research, development and manufacture of aerospace systems, including manned and unmanned aircraft; missiles, space launch vehicles and spacecraft; propulsion, guidance and control units for all the foregoing; and a variety of airborne and ground-based equipment essential to the operation of flight vehicles."

the Institute (March 25 through April 3, 1957) on the subject of automatic programming for NC. We defer a discussion of this course until Chapter 6, and only comment here that it served to emphasize the importance of advancing the art of programming for NC; it also provided an excellent platform for Ross to expound on his special purpose APT language, and served to draw him more closely into the workings of the SNC.

Following the January meeting in Dallas, Ross had many communications with SNC members and with their programming task force on matters of automatic programming and how he thought the AIA should proceed. Gradually he was able to get across the notion of a *systems approach*; whenever an opportunity presented itself, he would casually toss out the idea of a joint programming effort among AIA members and Servo Lab. This laid-back, soft-sell approach had a telling effect: at its April 25–26 meeting, the SNC gave formal approval to a plan in which MIT would lead the development of the first APT system based on MIT's research. This system would be known as the 2D-APT-II system: the 2D signified that it would be confined to two-dimensional surfaces and the Roman II meant that programming was to be accomplished through use of curves rather than by points or regions. Servo Lab was chosen as the project coordinator.

Although the 2D-APT-II effort was approved in principle at the April meeting, details of the language and how the joint project should be organized were left unresolved. However, Ross lost no time in getting himself and his Servo Lab group geared up for the work ahead. A one-week meeting of all programming personnel assigned to the project by the participating companies was convened at MIT on May 20, 1957, at which time Ross presented a detailed outline, not only of his ideas for 2D-APT-II but also an organizational structure for company assignments, reporting procedures, and a timetable for carrying out individual assignments. None the less, at the end of the May meeting, details of the 2D-APT-II language were still an open matter; the general concept of a system-oriented structure had been accepted but details within that structure were absent. The problem of settling on a language was exacerbated by the fact that Ross had to deal with both the programmers assigned to the project by SNC and with SNC members themselves. Thus, in preparation for the May meeting with SNC, Ross had solicited from all concerned their recommendations on matters pertaining to language vocabulary and their thoughts on definitions of contours such as lines, circles, general curves, and so forth. He was not enamoured by what he received; suggestions leaned more toward the close-ended Siegel approach than toward his own three-dimensional, generalized approach. Ross therefore concluded that he must take initiative and create the language himself, promptly. According to his own account, he did so over the weekend of May 25–26, 1957, and it was published as Memorandum No. 2 on June 14, 1957 under the title: "A Proposed Basic Language for the 2D-APT-II" (4.13). Although compromises were made as the design evolved, Ross reflected in his APT history: "The basic concept of the language has [with]-stood the test of time very well." The 2D-APT-II project proceeded within the framework of language set forth by Ross in his June 14 memorandum.

Nine aircraft companies, plus IBM at White Plains, New York and MIT, participated in the joint effort of Phase I. Since some of the aircraft companies contributed programmers from at least two geographic locations, a total of fourteen entities was represented in all. Their names and assignments are shown in Fig. 4.7. Note that 2D-APT-II evolved in two phases, with MIT as project coordinator for Phase I and North American Aviation as AIA's first-year coordinator for the upgraded version designated 2D-APT-II Phase II. The original organizations in the joint project are denoted by an asterisk. Figure 4.7 also reveals the modular structure of the APT system; its major parts consist of the *input translator*; the *pre-processor*; the

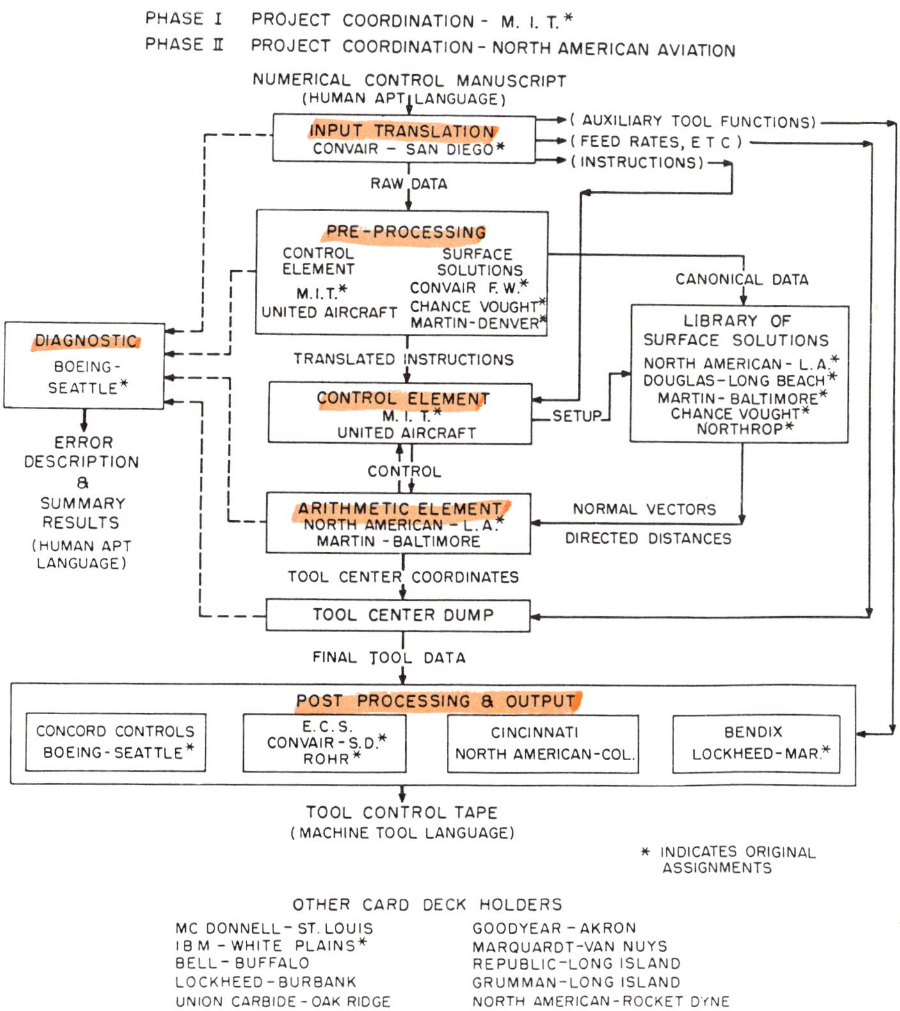

FIG. 4.7. APT programming assignments as of February 1959.

control element and the *arithmetic element*, with their library of surface solutions; and the *post-processor*, to provide an output match to the four different digital directors being manufactured at that time by Concord Control, Electronic Control Systems, Cincinnati Milling Machine, and Bendix. Another important part of the APT programming system was also the set of diagnostic programs shown by the box at the left in Fig. 4.7. Note that all the principal elements of the system interface with the diagnostic routines that had been designed to spot errors at each subsystem level of the overall program.

As might be expected, start-up of a major joint effort among so many disparate, geographically dispersed organizations was slow. Programmers had to be broken loose from other jobs, and as they blended into their new work, each had to capture not only the spirit of Ross's language philosophy but develop it so that it would be consistent with other component packages with which it must interface. A factor in favor of the group effort was that all the participating aircraft companies had IBM 704 computers; thus a single type of machine could be used as the basic one for which all programs would be written. Although this compatibility was very helpful it did delay Servo Lab's contribution somewhat, since MIT lagged behind industry in getting its 704 into operation. Until early 1958, the Servo Lab group had to use the one-of-a-kind Whirlwind machine as its computational vehicle.

Perhaps the salient factors that worked strongly in favor of the joint effort were the competence and enthusiasm of the participants and the cohesiveness with which they worked as a group once start-up roughnesses were smoothed over. Another positive factor was the close lines of communication that were maintained among all participants and the oversight groups—the AIA's Manufacturing Committee and its Subcommittee on Numerical Control. A hub-and-spoke type reporting network, in which information flowed inward from the field to Servo Lab and then outward to all participating companies, served to keep everyone informed and up-to-date. Each outgoing memorandum contained answers to queries raised by individual programmers, comments by programmers on one another's work, and critiques and suggestions by Servo Lab staff. It has often been said that a camel is a horse designed by a committee, but thanks to all the favorable factors cited above and to the powerful workhorse in the driver's seat, Doug Ross, it did not turn out that way: the 2D-APT-II system was well thought out and effective, just what was needed to put the whole NC concept on a firm and lasting foundation.

The Phase I 2D-APT-II system was ready for field testing at the participating companies' facilities within a year—specifically, on April 30, 1958. Field tests turned up the usual number of bugs that had to be corrected, and the corrections had to be coordinated so that all participants would have the most up-to-date version. Coordination of corrections and changes during the debugging phase was turned over to O. Dale Smith at North American Aviation. This was Smith's "warm-up exercise" in preparation for his taking over total coordinating responsibility for the next round of APT upgrading

—the 2D-APT-II Phase II system. While Phase I was being debugged, Servo Lab turned its attention to documentation, distributed at various times during 1959, and consisting of six volumes totalling some 517 pages (see bibliography).

Once the Phase I system was completed, the plan called for the AIA to assume field responsibility for its further development and for Servo Lab to serve only in an advisory capacity. Programmers for Phase II reported to the Project Group of AIA's Numerical Control Panel. From March of 1958 through October of 1959, they were under the direction of O. Dale Smith, followed by Len Austin at McDonnell from April of 1959 through October of 1960. Late in the fall of 1960, the AIA further asserted its commitment to APT by forming the APT Central Project. Fourteen programmers were assigned by their companies to work under the direction of E. A. Bates (General Dynamics, Convair). This Central Project Group, stationed in San Diego, began further APT development work in January, 1961, and completed their assignment in December, 1961.

Although Servo Lab was not supposed to have a formal programming assignment for Phase II, Ross's expertise was frequently needed to guide and critique field tests and to help out in the upgrading and refinement process. In fact, Servo Lab collaborated with the AIA at various intensity levels until mid-1962. A particularly troublesome component was the arithmetic element (ARELEM). As originally designed, it proved to be erratic and unreliable. To surmount this stumbling block, the AIA requested Servo Lab to reexamine ARELEM, with the result that Ross and his group worked out two new versions, one based on straight-line approximations and another on curvilinear approximations. The first version was incorporated into AIA's Phase II system.

While the ARELEM was being redesigned, part of the Servo Lab programming staff turned its attention to research on APT-III, programming by regions, and some experimental metal cutting was actually performed using this technique. The results gave dramatic reductions in programming time. For example, the elliptical paraboloid shown in Fig. 4.8 required only six additional programming commands beyond those needed to describe the two-dimensional elliptical base in order to contour the third dimension. The program automatically generated a spiral-cut pattern, starting at the base and working upward, in accordance with APT instructions. Sculpturing the same part with the 2D-APT-II system would have required a separate command for each cut vector around the part and one indexing command per trip around. For the part shown in Fig. 4.8, a rather coarse cut was purposely chosen in order to highlight the cut vectors. Although APT-III greatly reduced the programmer's job, it had limitations; it was not always possible to break complicated parts into three-dimensional regions that could be programmed by the regional technique. Programming by regions worked well when parts could be described in terms of conic sections, but that was not always possible. For that reason, APT-III never caught on.

As APT attained firm footing as a programming language for machine tools, it became evident to the AIA that an outside organization would be required to maintain the language, upgrade it, and handle the ever-increasing number of requests that were arriving at AIA headquarters for up-to-date decks of cards that constituted the system. To relieve itself of these burdens, the AIA decided to invite proposals from organizations interested in being the focal point for a long-range APT development and maintenance project. The selected organization would be supported through annual subscriptions from user companies. Although we at Servo Lab were already

FIG. 4.8. A rough elliptical parabolical cut using APT-III.

moving into a follow-on project—computer-aided design and computer graphics—we did submit a proposal. In retrospect, it was fortuitous that the AIA chose to set up the Illinois Institute of Technology Research Institute (IITRI), formerly the Armour Research Foundation, rather than designate Servo Lab as the long-range agency. Had Servo Lab been selected, some of our research staff would have been confined to APT affairs and thus diverted from the research challenges that our computer-aided design project was presenting. Once the long-range APT organization was selected, future maintenance and upgrading of APT were assured.

The AIA retained control of APT through 1962, at which time IITRI was able to announce completion of a working version of its APT-III, not to be confused with Servo Lab's programming by regions, which was also called APT-III.*

Once APT-III was announced, and IITRI was firmly established as an APT maintenance and development center, the AIA withdrew, putting APT's future in the hands of IITRI. Included in the AIA–IITRI public announcement of the completion of APT-III was a photograph of the part shown in Fig. 4.9. The AIA commented on the photograph as follows:

> The part in the above photograph emphasizes the trend to integrally machined structures to obtain greatest strength and product reliability at lower cost. This machined aluminum-alloy forging replaces an assembly of over 250 sheet metal parts and graphically illustrates complex contouring on both internal and external surface dimensions previously attainable only by the use of several machine tools and many more tedious man-hours with corresponding higher cost. Total tooling costs and the number of tooling items were greatly reduced by the use of APT-III application to tooling manufacture and part production.

Very early in the APT-development program its power as a programming concept became evident to the closely knit group of programmers and programmer managers who were involved in the joint AIA—MIT effort, but during the period 1956–57 while it was under development, the system was relatively unknown beyond those groups. Hence, as the 2D-APT-II Phase I version neared completion, and there was confidence that an important

*APT nomenclature can be confusing. As already explained, 2D-APT-II Phase I referred to the version developed jointly by MIT and AIA under MIT's supervision. Although confined to the cutting of two-dimensional surfaces described by planar geometric curves, the two-dimensional constraint was not an inherent limitation of the system itself, only a constraint agreed upon in order to get a working system into operation in a relatively short time. APT-II was fully capable of programming three-dimensional cuts. In order to differentiate between the 2D-APT-II Phase I and the upgraded version of it, developed largely by the joint efforts of the aircraft companies themselves, the label "Phase II" was substituted. When AIA's Central Project Group took over further programming responsibility in 1960, it produced a rudimentary version of what became known as APT-III, which was passed on to IITRI, where it was further developed and turned into a fully operational APT-III system.

IITRI also developed a subset of programs that ran on smaller computers; this version was called ADAPT, not to be confused with still another version of APT, developed in Germany, also called ADAPT.

milestone had been reached, there was also an awareness that the time had come to inform the professional public about research results and to excite broad interest in the language. During 1958, Ross presented papers at meetings of the American Association of Arts and Science and the American Society of Mechanical Engineers, and plans were made for a press conference to be held at MIT in early 1959. The conference, held on February 25, 1959, was sponsored jointly by MIT, the AIA, and the Air Force, and is commented on in detail in Chapter 6.

Ross's publications and the press conference served to orient the thinking of the program users' community in the direction of APT and to get it established as the accepted language throughout the aircraft industry—so much so that in 1974 the APT language became the United States standard for programming numerically controlled metal-cutting tools. Although originally conceived as a language for programming continuous-path numerically controlled machine tools in the aircraft industry, its use gradually broadened to a variety of industries employing numerically controlled processes. In accordance with the American National Standards Institute (ANSI) requirement that its standards be reviewed every five years, APT is

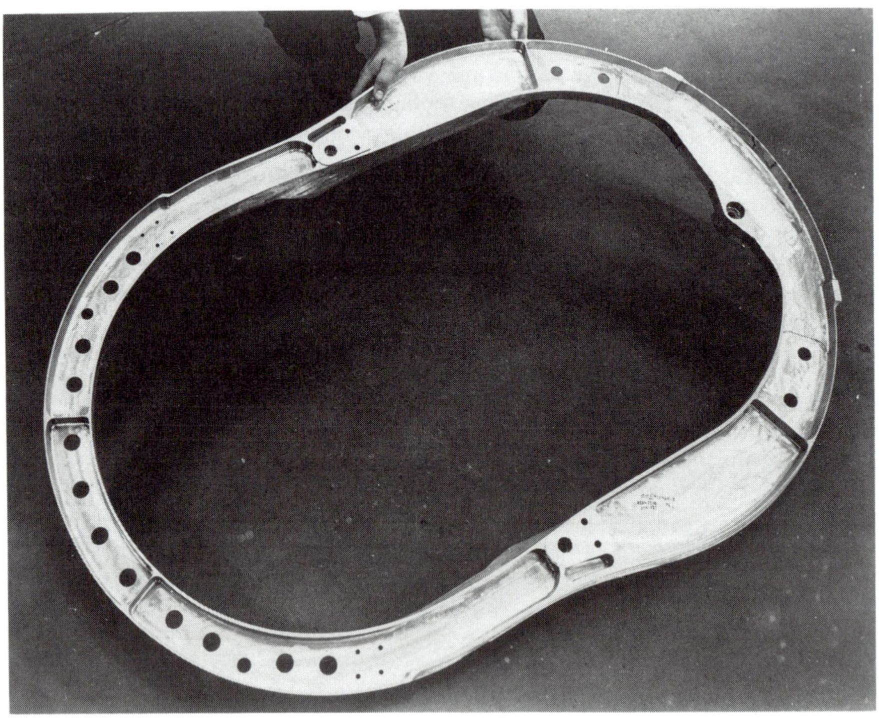

FIG. 4.9. An aluminum alloy forging machined through use of IITRI's APT-III system of programming. Courtesy of the Aerospace Industries Association.

currently in its final stages of another five-year updating; it will appear shortly as ANSI Standard X3.7-87.

APT was nominated as an international standard in 1967. The first presentation was made to the International Standards Organization's (ISO's) committee on special languages by ANSI's subcommittee on APT standards at an ISO meeting held in Paris, November 6–10, 1967. The arguments set forth by the US delegation were strong: APT was totally nonproprietary; as of 1967, thousands of programmers were using it throughout the United States, Europe and Japan; it was being maintained and continually upgraded; and its modular features made it relatively easy to expand and modify. As might be expected, establishing an international standard is a long and tedious process. None the less, the ISO committee was able to work out a plan that was ultimately acceptable to all participating nations, with the result that APT became an international standard in 1978. To achieve an agreement, however, ISO had to depart somewhat from the US standard. It established three subsets of standards within APT, one for the input-language translator, another for the cutter-location file, and a third for the output form of APT.

Personnel, Contractual, and Other Matters

The APT project started during the summer of 1956 with Ross as the sole full-time staff member. As previously noted, the immediate job at hand was to develop a group of programming specialists around Ross as the technical leader. Over the ensuing twelve months, manpower expanded gradually, the average rate of effort over that period being 4.7 full-time-equivalent staff members. The average head count was higher than 4.7, however, since it was our practice to give recognition to the time graduate students spent attending classes by counting their project assignments as a fraction of a full-time commitment.

Manpower continued to grow steadily during the next two years; the average full-time equivalent during the 1957–58 fiscal year was 6.3 and during the next fiscal year the number grew to 7.6. It would, of course, have been impossible for MIT to accomplish all that transpired over the three-year period in the APT development without the many tens of man-years of programming effort contributed by the aircraft industry. At a meeting of APT programmers held at MIT from April 1 through May 1, 1959, 22 persons from the aircraft companies that were actively engaged in the APT cooperative project were in attendance.

Our Air Force contract, AF33 (038)-24007, with its starting date of February 8, 1951, came to a close on November 30, 1959. Over its lifespan of eight years and ten months, total expenditures, including those on the original contract and its four supplements, amounted to $1,051,750—an average yearly amount of $119,000. We defer comments on the return the nation received on its investment until Chapter 7.

In 1959, we changed the name of our Laboratory from Servomechanisms

Laboratory to Electronic Systems Laboratory. The new name reflected the many changes in the nature of the research that had occurred during the first twenty years of the Laboratory's existence. We concluded that the time had come for a name change when a number of our incoming graduate students would comment: "This is a great laboratory, the projects are excellent, but what have they to do with servomechanisms?" In keeping with our desire to be at the cutting edge of new technology, we had abandoned the mature technology of servos in favor of new, challenging areas involving broader aspects of electronic components and systems—solid-state radar components, chemical-process control, controls for underwater vehicles, digital flight simulators, and biomedical problems, to name a few.

5

THE LATER YEARS

The Origins of Computer-Aided Design Research

The later years of our project spanned a ten-year time period from 1960 through January 1970, during which time the underlying research theme was computer-aided design and interactive computer graphics. Because of the Laboratory's forefront position in the development of digital computers and automatic control, and the successful merging of these technologies into a real-time control system for machine tools, we were called upon frequently for presentations on these subjects. There was intense interest in the 1950s, not only about details of these technologies but also about their impacts on the future of manufacturing from the viewpoints of productivity improvement, job formation and job elimination. *Automation*, a newly coined word, was being bandied about by some as our economic salvation and by others as a socioeconomic disaster.

The many talks we gave on the subject of automation served to crystallize our thinking about the potential impact of numerical control on the mechanical design and production process as a whole. We realized that machining a part was only the final stage of a multistage process that began with a functional requirement for the part and progressed through several steps that included the setting of specifications, design execution, mechanical drawings, detail drawings, machine operator's instructions and finally machining. The mechanical drawing came under close scrutiny. Obviously the conventional blueprint was an information storage medium; it stored on paper, in codified, graphical and symbolic language, all the information needed to machine the required part, and the language of the blueprint was the language of both the draftsman and the machinist. However, when machining was performed automatically by numerical control rather than by manual methods, a translation from the language of the draftsman to that of the machine tool was required.

This translation requirement naturally raised the question of whether the traditional drawing contained the right kind of information when it was known in advance that machining would be under numerical control. Perhaps the traditional blueprint language, so well understood by the mechanical designer, draftsman and machinist, should be altered or expanded to take into account the presence of a part programmer in the design process.

Inevitably the deeper question arose as to whether or not a mechanical drawing would be needed at all in a computer-oriented design setting. When a design was stored on paper, a graphical language was required to express and store the information, but in the computer-oriented case, why not use a numerical language to express and store the same information in the memory of the computer, allow the computer to generate and display a picture of the piece, and dispense with the blueprint? This approach should permit rapid refinement of the design.

Our thinking actually went beyond the mechanical drawing; it penetrated into the very design process itself. We envisioned the possibility of the designer working back and forth with a computer as a design tool, rather than with a slide rule or a desk calculator. We saw the synthesis process being executed by a designer who would sketch trial designs on a cathode-ray tube face rather than on scratch paper. It was against this backdrop of thinking that our plans for a computer-aided design project began to solidify, even before the APT project had come to an end. Several points became clear as we probed into the problem more deeply: if we were to engage in research related to mechanical design, mechanical design engineers should be involved; if the designer were to execute his designs with the aid of a digital computer, some kind of workstation would be required that would substitute for drawing board, slide rule, desk calculator, pencil and paper. We reasoned intuitively that in order to achieve a mechanical design with the computer as a design aid, an elaborate set of programs would be required, and that these should be developed with both the creative process and NC machining in mind.

We had no illusions about the magnitude of an undertaking of this kind and the barriers that would have to be overcome. Our APT experience had taught us that special purpose computer language development was a nontrivial task, and that our sponsor, while enthusiastic about the work, would probably not underwrite research that appeared at the outset to be even more formidable and riskier than the APT project. Fortunately, events were taking place in other parts of MIT that would eventually supplement the support we would receive from our sponsor. As the APT project was coming to a close, important things were happening within MIT's computational facilities. These matters are discussed in detail later in this chapter. Suffice it to state here that our Computation Center, which had been providing Institute-wide services without cost, began to take steps toward a fee-for-service structure that would impose charges in accordance with the amount of time an individual utilized the machine. Also under intensive investigation was the idea of an online timesharing service that would allow users to interact remotely with the central computer by means of keyboard-type terminals (teletypewriters) and phone lines. Thus, there was emerging on campus a modern, mainframe computer environment which could effectively serve our needs for developing a large programming system for design purposes, and also an environment that would undoubtedly be compatible with a research project directed toward online graphical displays built around cathode-ray tubes as well as teletypewriters. The fee-for-service plan caused us some anxiety, however, because from the very beginning of our project it had been unnecessary to include a computer usage fee in our budgets to the Air Force. Both the Whirlwind and the Computation Center's machines were readily available for on-campus research at no cost to the project. Thus, new Servo Lab research proposals would have to include another expense item that would need to be sold to the sponsor. Some detailed explanations were necessary, but as noted later, we were able to gain acceptance of the fee-for-service requirement.

We began to lay groundwork for a computer-aided design contract with our Air Force sponsor almost a year in advance of the expiration date of the APT contract. In a monthly summary letter to the sponsor, we reported that: ". . . in January [1959] an informal meeting was held with personnel of the Design Division of MIT's Mechanical Engineering Department. They displayed considerable interest in automatic design [and] plans are to continue discussions in this area. . . ." Two months later, we were able to report that Professor Laurel Lewis, a visiting faculty member in our Laboratory from the University of Washington, and Dwight Baumann, an Instructor in MIT's Mechanical Engineering Department, were actively studying the feasibility of applying automatic programming to the design process and were to present their findings to a larger group of Laboratory and Mechanical Engineering Department staff within a month or so. As a vehicle for open discussions, a seminar series was established, and the first seminar was held in April, 1959. In attendance were four Laboratory representatives, Professors Reintjes and Lewis and Messrs Ross and Clements, and seven representatives from Mechanical Engineering, Professors R. W. Mann, J. L. Shearer, J. B. Reswick, S. A. Coons, H. Richardson, D. Lent, and D. Baumann. Part of the record of this seminar stated: ". . . there seems to be a great deal of interest in the possibilities for [research] in this area," and also expressed the opinion that the seminars should continue.

At the second seminar, held on May 8, 1959, there was much mind-flexing and brainstorming as the group got down to serious thinking about what actually takes place in the design process and what roles computers might play in it. Gradually a scenario developed in which the designer would be seated at a computer console, drawing a sketch of the device he had in mind on the screen of a cathode-ray tube with the aid of his "light pen" or other input device. It was envisioned that he would be able to modify the sketch at will and command the computer to redraw the device in accordance with his updated ideas. The computer also would be able to transform the designer's rough light-pen motions into highly refined pictorial drawings and simultaneously produce digitized data that could be used to drive a numerically controlled drafting machine, or simply to do further processing in preparation for NC machining. Other operations were also envisioned for the computer. It could be instructed to carry out certain numerical analyses —for example, stress analysis—which would be essential in the design process. Most importantly, though, it was concluded that the computer should *complement* the work of the designer during the whole process; at some times, it would be the designer's slave, at others it would alert the designer to impossible requirements or constraints being imposed.

Two kinds of engineering design were identified: *repetitive* design and *creative* design. In repetitive design, the problem is completely defined analytically, and hence can be executed directly by the computer; in creative design, the design is only vaguely defined and requires a series of judgments from the designer, and perhaps even one or more modifications of specifications as it proceeds. Examples of equipment or devices being repetitively and automatically executed, even in those days, were electric motors, transformers, turbines, stressed structures, and so forth. In these applications, all steps in the design process are known, and can be expressed in a form that can be executed by the computer without human intervention. In other words, the design steps exist in handbook form, and the designer has only to apply specific numbers to the general variables, after which the computer generates a specific version of the device within its class.

It was concluded very early in our discussions with our mechanical engineering colleagues that our project should not concern itself with repetitive automated design; rather, we should address the more difficult problem of bringing computers into the creative design process. Although it was unclear what role, or roles, the machine was to play in the creative process, it was still reasonable to expect that it would be unable to carry out a design completely unaided by the designer, since at the outset only the functional requirements would be known; the *goal* of the work would be a design to meet those requirements. Hence, it was envisioned that the computer would serve as a *design tool*, not as a replacement for the designer. As we continued our deliberations, we eventually abandoned the term automatic design altogether and adopted the phrase *computer-aided design*. This point may now seem trivial, but at the time it required considerable effort to convince our listeners that when we spoke about computer-aided design, we were taking a step forward, not backward, that we were using the term in the context of the really tough problem of creative design, not in relation to the much simpler problem of repetitive automatic design.

The seminars we held with our colleagues helped us to focus attention on

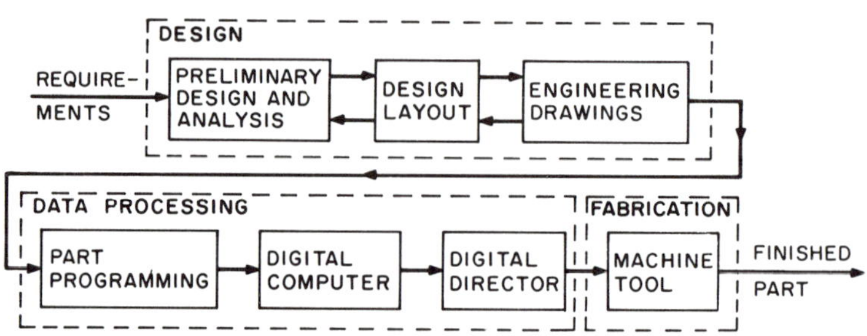

FIG. 5.1. The overall design and machining process, with machining by numerical control.

the complete manufacturing process and to describe it in computer terms. Thus, in Fig. 5.1, the process is divided into three major stages: *design, data processing and fabrication.* The design stage begins with a set of functional requirements and proceeds through three steps: (1) a preliminary (trial) design is made, analyzed to determine its ability to meet requirements, and the design is altered, as necessary; after which (2) a final design layout is produced; and finally (3) a set of drawings is prepared. At the time of our initial studies, the design and drawing operations, of course, were human operations. As indicated in Fig. 5.1, the second stage, data processing, began with the engineering drawings that a part programmer used to translate from a graphical description of the part to a numerical description. A digital computer would next be used to transform the programmer's numerical data into a form that could be understood by the digital director that drives the machine tool. The final stage of this three-stage process, fabrication, is then carried out by the numerically controlled machine tool.

Very early in our discussions of how to move the manufacturing process into the domain of the computer specialist, we attempted to describe the manufacturing process in terms of the computer specialist's language, with emphasis on words and phrases such as *information*, *information generation*, *information storage*, and *information processing*. Figure 5.2 illustrates the process from this viewpoint. (The diagram is reproduced in Fig. 5.3 for clar-

FIG. 5.2. Reintjes (center) outlines the role of computers in manufacturing processes, with Ward (left) and Professor Newton (*circa* 1959).

ity.) Thus, in Fig. 5.3, executing a design is basically an information generation operation; translating design information into a set of instructions that can be understood by the NC machine tool is information processing; and employing that information for machine tool motion control is information utilization. Since the whole process cannot be executed in real time, provision must be made to store information along the way in an appropriate form: in graphical form at the blueprint stage; as computer instruction words, on paper, at the part-programmer stage; and in numerical form as information passes through the general-purpose-computer and special-purpose-computer (digital-director) stages.

Even before our first proposal for a research project in computer-aided design had been put together, our staff members recognized that executing creative designs is as much an art as it is a science. Not only are there no rigid sets of rules and design steps, it is often possible to satisfy a given set of requirements with more than one physical embodiment. Individual designers have their own design tastes and procedures, which, while producing a part that meets the requirements, will also often differ in detail from designer to designer. From the computer-aided design viewpoint, this was both good and bad. The good news was that if the designer knew that machining would be done under numerical control, he might be able to introduce design features that could take advantage of three-dimensional automatic machining, with possible savings in material and machining costs. The bad news was that an elaborate set of computer programs would be needed in any computer-aided design environment to enable the designer to operate effectively. In any event, it was clear that the computer-aided design environment should be a *man–machine environment*: that is, the designer should be able to work back and forth freely with the computer and its associated workstation as the design progressed through its various stages (5–1, 5–2).

When it came time to submit a proposal to our sponsor (still the Manufacturing Methods Branch of the Air Materiel Command), both our Laboratory group and the Mechanical Engineering group knew what the general

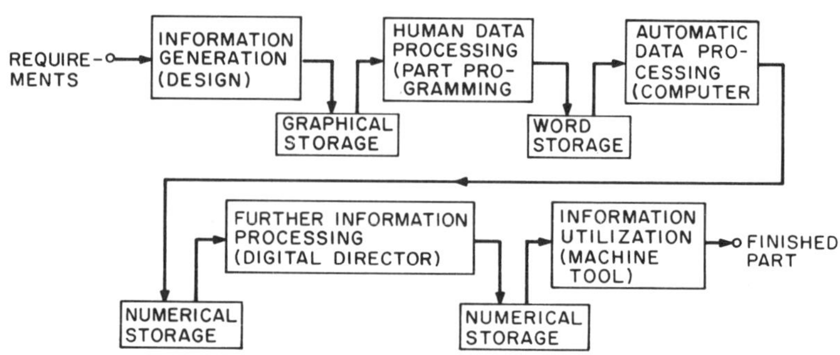

FIG. 5.3. Representation of the manufacturing process, with emphasis on information generation, storage, and processing.

nature of the tasks should be: namely, a critical analysis of what roles the computer should play for creative design; the development of a system of programs for the creative design process; and an investigation of requirements for a workstation that would serve as the interface between the designer and the design tool—the general purpose computer. A proposal along those lines was submitted to the Air Force on June 16, 1959, outlining a one-year effort for a cost of $223,000. It was entitled: "Investigations in Computer-Aided Design for Numerically Controlled Manufacturing Processes." The study's broad objective was "to devise techniques for automatic programming of design information that are effective, economical and adaptable for use by the aircraft industry in conjunction with the APT System and numerically controlled manufacturing."

Five tasks were proposed. The first pertained to the development of computer languages and structures convenient for human usage and suitable for manipulating design information. The second was for "the development of means for pictorial and symbolic communication between the human and the computer, using oscilloscope displays, light pens, teletypewriter, and switch inputs to the computer." The third task centered on the mechanical design process itself and called for a "study of the mechanical design process," as it was then practiced, but for the express purpose of determining in which of the design functions computers could be "of direct benefit." The remaining two tasks were in the nature of project support and provided for continued technical guidance to the AIA–APT project (which, as it turned out, required more effort than originally estimated) and for maintenance of the MIT experimental NC machine tool.

We encountered no difficulty in getting the proposal funded, even though it was necessary for the first time to include a line item for computer usage. We had kept our Air Force project engineer, Mr Webster, fully informed about our preliminary discussions on computer-aided design and our desire to push in that direction, and also about the Institute's new, upcoming policy to impose a fee for use of its computer facilities. The proposal culminated in Air Force contract AF33 (600)-40604, dated December 1, 1959, for one-year funding in the amount of $223,000; included was a line item for 104 hours of computer use at $200 per hour.

While the computer usage fee issue should not be overemphasized, remember that for nine years we had been working with this sponsor without requiring payments for computer time. Over the four-year APT development period, we had made heavy use of Whirlwind I, and to a lesser extent of the Institute's IBM 704 facility, at no cost to the Air Force. Naturally, we were fearful that this sudden inclusion of a $20,800 computer-usage item in the new proposal might shake our contracting officers to such an extent that we would be without research funds while awaiting a high-level government policy decision. Fortunately, either because of the persuasiveness of our Air Force project engineer or because government red tape was less red in those days, or more likely because both factors were true, our fears proved ill-founded: no gap occurred in research funding.

The Years 1960–1965

The computer-aided design (CAD) project got off to a slow start, partly because the Laboratory's commitment to redesign APT's arithmetic element (ARELEM) was consuming more time and manpower than anticipated, and partly because both our Laboratory and the Mechanical Engineering Department staff were feeling their way cautiously as they planned specific research efforts within the scope of the research contract.

[An interesting sidelight, which also contributed to the delay in research, occurred in 1960. In conjunction with MIT's Centennial Celebration, the Columbia Broadcasting Company planned to put together six one-hour television programs on various "futuristic" scientific topics under the title of the "Tomorrow" series. The second program would be concerned with "Machines that Think." As a result of discussions at the Institute, one of the applications given serious consideration for this segment was computer-aided design. At first it was thought that a simple mechanical device—a wheelbarrow for example—could be designed in real time by a computer during the show, but lack of an adequate programming system and a visual display terminal led to the abandonment of this idea. Several other possibilities were considered and discarded for one reason or another, until finally the frustrated CBS producer suggested that the plots of TV Westerns were so stylized that a computer might be able to synthesize a short TV Western skit or play. Ross liked that idea because he thought it would not only be an illustration of artificial intelligence, but it would also show that intelligent behavior is basically rule-obeying behavior. Script-writing rules, when properly contrived and organized, should yield believable action. Upon further thought, Ross came up with an idea for a computer program that would allow the machine to create a series of skits, all on the same theme but with different random scenarios. The theme would be a TV Western with two characters, a sheriff and a robber, and the setting would be a hideout shack with a table, a money bag (the loot), a whiskey bottle, a jigger, and so forth. The plan was to give the computer a capability to generate a series of skits of its own choice, each with a different version of a central theme, and to have the skit acted out in pantomime by two actors in accordance with one of the machine-generated scripts.The computer program had to be written so that all scripts obeyed "intelligent rules," meaning that, as alternative versions were synthesized, the result would make sense in terms of human behavior. An interesting feature of the program was the inclusion of an "inebriation factor," which caused the rule structure to be aborted and restored randomly, as though the affected actor had forgotten what had been his goal. This erratic behavior depended on how often the actor went to the whiskey bottle. Thus, proper drunken (nonintelligent) behavior resulted.

The rationalization for the whole effort was to demonstrate that the machine could achieve some measure of creativity in a situation which was nonmathematical, and that the product of the machine's creativity could be described in terms laymen could understand. When the program was first run, it contained a bug: it created one skit in which the robber pulled his gun and put it in the sheriff's holster. The intent had been to have the actor respond to: "Put the gun in the holster," but since Ross and his colleague had neglected to specify *which* holster, an allowable response to the rule was to put the gun in the other fellow's holster. Of course, this sequence did not occur in all possible versions of the script—alternative plots were randomly generated. When completed, the program, called SAGA II, satisfied CBS to a tee. It was aired during a one-hour segment of their "Tomorrow" series (three skits were presented, including the one with the glitch). For the record, CBS underwrote all costs of its preparation.. Many details of SAGA II are presented in John Pfeiffer's book, *The Thinking Machine* (5–3).]

Working closely with Ross on the redesign of ARELEM was Clarence Feldmann, who was given primary responsibility for the redesign and reprogramming. Feldmann devoted full time to the job, including a substantial amount of time at San Diego, where the AIA's Central APT programming effort was headquartered. It was evident from a progress report he gave at an APT programmers' meeting in early March, 1961 that he would need several more months to produce a working version of the new arithmetic element, and it would be necessary for him to spend at least two months in San Diego to coordinate the programmers' work with the new ARELEM. This necessity further delayed application of experienced manpower to the CAD project.

The March 1961 programmers' meeting brought forth many favorable comments about APT and many success stories about its use. Nevertheless, frustrations also were expressed, largely because of program bugs or basic deficiencies in the very complex system. On balance, though, the meeting was upbeat and revealed steady progress toward a highly versatile, unified, reliable APT system. The programmers' enthusiasm for APT was self-evident; when programming glitches were brought into the open, a programmer from one or another aircraft company would come forward and volunteer to correct the problem. At the March meeting, one of our former staff members, Sam Matsa, who had left us to join IBM, described his version of an alternative automatic programming system for NC. Labelled AUTOPROMT, Matsa's system was, in essence, a composite of certain APT-III features and his own ideas. Although briefly in competition with AIA's APT system in the early 1960s, AUTOPROMT was destined to have a short lifespan, since IBM chose neither to continue its development nor to maintain it. At one point during the March meeting it was concluded that some thought should be given to incorporating certain AUTOPROMT features into APT, but this never happened.

Meanwhile, Dwight Baumann and Professors Mann and Coons began to focus on specific topics for their first studies of CAD. Both Mann and Coons brought a wealth of expertise to the digital design project. As Mann explains it, design has occupied a major part of his life from the beginning of his technical high school days where emphasis was placed on mechanical drafting, descriptive geometry, and graphical representations of mechanical devices. His interest in design continued during his undergraduate and graduate years at MIT, so when it came time for him to join the Mechanical Engineering Department's faculty, his obvious niche was in the Department's Design Division. Even before taking up that post, Mann had utilized computers in design applications, particularly analog computers. He was also aware of the potential of digital computation through his contacts with two Servo Lab staff members associated with the Whirlwind project—his friend (and later his wife) Margaret Florencourt, and his undergraduate roommate, Kenneth Olsen, now president of Digital Equipment Corporation. As a result of several faculty changes that occurred in rapid succession in the Design Division, Mann soon found himself in charge of that segment of his

Robert W. Mann

Department's curriculum at the time the CAD project was getting under way.

Working closely with Mann was Steven A. Coons, also an experienced designer. During World War II, Coons had spent a major amount of time at the Vought Aircraft Company, designing and fairing aerodynamic surfaces, for which calculations were mainly carried out through the use of mechanical calculators. Thus he knew firsthand about the tedium of that job. This experience, as well as his fluency in projection geometry and his uncanny knack for solving analytical problems by geometric means, made him a valuable assistant to Mann when Mann assumed the leadership of Mechanical Engineering's Design Division. It was the team of Mann and Coons that spearheaded the Division's work on the CAD project, with Mann giving Coons continual encouragement to take a strong leadership role.

At the outset of the project, Mann and Coons concluded that in order for them to understand just how digital computers might aid in the design process, they would first have to know how designers would communicate with the machine, and how analysis should be handled in the machine environment. Another topic that attracted their attention was the division of work between designer and computer during the design process. Four graduate students were recruited to examine these issues: R. P. Parmelee chose to investigate the important area of stress analysis in the machine environment; J. A. Verderber and L. E. Johnston undertook investigations of the man-machine interface problem; and J. D. Purvis, Jr, studied how a general purpose computer might serve as a standard-parts selection facility.

All four students penetrated to the nub of the issues related to their topics and proposed machine specifications that would contribute constructively to

Steven A. Coons

the design process. With respect to stress analysis, Parmelee tackled the central question: What features should a computational facility have in order to assist the designer-analyst in a way that will *elevate his confidence in his design* and *do the job with some time and cost savings*? Parmelee, Mann and Coons saw excellent possibilities for the computer in stress analysis, not because it would contribute to the theory (the mathematics of stress analysis and the underlying theory were well known), but.because it was hoped that, through use of a computer, many more checkpoints on stress could be analyzed quickly and economically. Since hand calculation was long, tedious, and often only approximate, the conventional design tendency was to limit the number of calculations to locations deemed critical in light of the designer's judgment and experience. By the time Parmelee was ready to submit his thesis document (in partial satisfaction of the requirements for the MS degree (5–4)), he had concluded that a general requirement for the computer was that it should have a mathematical problem-solving capability that would be easy for the designer to use, it should perform the stress calculations, and it should be able to display the results. Parmelee also saw the need for an interactive capability between designer and computer, and the need for a programming system that would provide flexibility during the design process. In other words, the designer should be able to issue instructions in English-like terms; have the freedom to change parameters at will and immediately see the resulting changes in stress; and have the option to change the *mode* of stress analysis—bending, tensile, torsion, and so forth. Computer programs written to permit all these operations would have to be complex indeed, but more importantly they would have to be compatible with the many others that would be required in a full-blown CAD programming system. As we shall see later, the job of putting such a system together fell to Ross and his associates.

Meanwhile, another Mechanical Engineering graduate student, P. T. Bennett, joined the project, concentrating on the development of a program for computer-aided analysis of beam stresses. His investigation was aided by the availability of an interactive graphics workstation, discussed in a later section of this chapter, with the result that he actually was able to implement a machine-aided stress analysis program for beams and parts of structures that could be modeled as beams. Bennett chose a specific design problem to demonstrate his interactive computer program: namely, the adjustment of the dimensions of an I-beam to achieve a uniform equivalent stress in the web and flanges. His approach was to use design handbook formulas for which he wrote computer programs. Besides producing satisfactory results in just a few minutes for simple loadings, Bennett's programs were able to show that, through their use, a designer could extend his skills by modifying the cross-section of his sample beam at will and rapidly testing the effect of each modification. Because of this speedy response to changes, the designer would be able to do a more thorough job of optimizing the cross-sections. Also, since all components of the stresses were easily calculated each time a change was made, the designer could develop a history of results, as changes were made, to determine which loads were critical—particularly if applied one by one. As Bennett observed, such detailed analysis was not practical when calculations had to be carried out by hand.

For several years after completing his MS thesis, Parmelee continued to investigate computer-aided stress analysis. His doctoral thesis (5–5) addressed the much more complicated problem of three-dimensional stress analysis, and emphasized a machine-oriented methodology (employing finite-difference equations) that would give fast, accurate results at low cost. For two-dimensional analysis, Parmelee's time and cost measurements were in the range of from 30 to 200 seconds at a computer computational cost of between $5 and $15.

The second important class of problem in mechanical design that was investigated by the Mechanical Engineering Department's Design Division concerned computer storage of standard-parts information. This work was undertaken by Professor Mann and his graduate student, J. D. Purvis, Jr. A rather straightforward logic had put this topic high on Mann's list of priorities: designers frequently included predesigned items in their creative designs; since the process of looking up and selecting such items from vendors' catalogs was tedious and repetitious, the designer would be helped materially if this low-level information search-and-retrieval task could be relegated to the computer, thereby allowing more time for his creative tasks. The strategy would be as follows: designers would specify the properties of the predesigned item, after which the computer would make the proper selection from its stored catalog of parts.

Mann and Purvis's work (and later that of R. E. Starzec, Coons' graduate student) was certainly ahead of its time. Although a few industrial organizations were storing a limited amount of catalog-type information in computers in the 1960s, widespread computer storage of handbooks and catalogs was just not possible because of the enormous amount of information involved and the absence of low-cost equipment to handle it. It was estimated that the 1960 version of the *Handbook of Physics and Chemistry* alone would require approximately 10 megabytes of storage, about equal to the total capacity of a secondary magnetic disk store in those days, and the storage unit would be available on a rental basis for approximately $60,000 per year. In contrast, in 1988 5 gigabytes of hard disk storage can be purchased for about $60,000.

Initially, Purvis limited his study to a single class of predesigned item: fasteners. These included bolts, nuts, screws, rivets, pins, nails, and so forth. After further narrowing the list to threaded fasteners, he sought to answer three questions: what must a designer know about threaded fasteners in order to make his selection; how should this information be stored to ensure effective retrieval; and how should retrieval programs be structured in order to put the primary burden on the machine rather than on the designer? These basic questions underlie computer-aided *fact retrieval* in general, and are the ones information specialists still grapple with as they attempt to cram more facts into computer storage and gain faster access to them.

Both Purvis and Starzec made headway on the general fact retrieval problem, both in terms of gaining a better understanding of how creative designers specify and select predesigned components and what problems program system analysts face as they attempt to structure the data and devise computer programs that manipulate and retrieve it. It is safe to say that both students, as students of mechanical engineering, received an exciting exposure to the growing interdisciplinary nature of engineering through their investigations. The research of each bore a striking resemblance to projects being undertaken by their counterparts in electrical engineering and computer science who were investigating techniques for structuring and compacting large quantities of data in computer files and developing programs for data recovery.

Professor Coons was a strong contributor to the CAD project, not only through his project leadership and the guidance he gave his thesis students, but also through his creative work in the development of mathematical methods for describing three-dimensional surfaces of arbitrary shape. There was considerable activity in this area during the 1960s because of the attractiveness of computers for solving the representative mathematical equations (see, for example, reference 5–6). At the time of Coons' work, analytical techniques for describing already existing surfaces were well known, but information on how to synthesize surfaces from mathematical relationships was sparse. The nub of Coons approach was to construct a complex, arbitrary surface from elementary surfaces pieced together to form the desired major surface. He called the elementary surfaces "patches," and they subsequently became known in the literature as "Coons patches." In the Coons-patch approach to design, the designer would start with a single surface patch, define it analytically in accordance with the mathematical procedure developed by

Coons, and then combine it with other patches that had been similarly defined. Thus, a surface of arbitrary size and shape would be comprised of an array of patches (5–7). Coons' mathematical method of defining surface patches ensured continuity between adjacent patches, even when modifications were made to the curves that defined the edges of the elementary patch, but it did not permit alterations within the patch. In the latter case, the designer would have to start over with a smaller size patch.

The beauty of the Coons'-patch approach was that it was easy for the designer to use, a minimum amount of input information was required, and the designer could modify his surface either by changing the original input or by adding more constraints. Although the mathematics involved was simple, it was still carried out most effectively by means of the computer. Each patch was defined by four boundary curves of any kind—circles, second-degree curves, polynomials, or non-mathematically defined curves—and provisions were made for manipulating the mathematical expressions for the curves and their derivatives in order to ensure that contiguous patches would be continuous at their junctions.

Coons patches were particularly attractive in the design of structures having considerable subtlety and complexity. Coons' favorite examples were aircraft fuselages, automobile bodies, ship hulls, and racing yachts. Designs for such surfaces were carried out traditionally by graphical procedures which required laborious trial-and-error calculations to ensure adequate descriptions to the desired accuracy. As part of their joint MS thesis, M. L. Hamilton and A. D. Weiss demonstrated the practicality of the Coons-patch approach to surface synthesis. Their work, published as a Laboratory report (5–8), involved the development of an elaborate set of programs that enabled a designer to execute a preliminary design of a ship through use of computer-aided techniques. Three-dimensional hull surfaces assembled from Coons patches could be displayed on a cathode-ray-tube screen, altered as desired in a few seconds by typing parameter changes, and the hull rotated to any desired angle for viewing. Through the use of their set of computer program routines, Hamilton and Weiss were able to closely simulate several parameters of US destroyer-type DD692 and to calculate to a reasonable degree of accuracy its midships coefficient, prismatic coefficient, displacement, wetted-surface area, and center of buoyancy.

Related Events

To put the second major task of the CAD project, development of interactive workstations, into perspective, we first wish to discuss in detail several events and decisions related to computer service that occurred at the Institute during the late 1950s and early 1960s, and to show the impact these events had on our project, One matter concerned the great ferment in machine computation taking place on campus during that time period, and the volatile computer environment in which we had to work. The complexity of the computer-aided design problem was of such magnitude that the availability of a large-scale general purpose computer was mandatory. However, finding a machine that would be available to us over the life of the project proved elusive. The Whirlwind computer which had served us so well during APT development was phased out for general computational purposes in 1958, making it necessary for us to turn to the Institute's Computation Center for service on its IBM 709 batch processing machine.* Service at the Center was often very slow. The demand from all parts of the Institute was increasing rapidly, but under MIT's agreement with IBM, only one eight-hour shift per day was available for Institute jobs. (The second shift was assigned to other regional universities and the third shift to IBM.) As a result, by the early 1960s, the worst-case turnaround time for processing a job could be as long as one or two days. A further complication arose because the Computation Center was not only providing service to the community, it was simultaneously undertaking its own research program on online timesharing techniques. In effect, therefore, in working toward its long-term goal of making computation more effective, the Center was competing for machine time with the community of users it was intended to serve. In an effort to accommodate everyone's interests, in 1962 the Center replaced its vacuum-tube 709 with the transistorized 7090, an upgraded version. Although a faster machine, the 7090 could still barely keep ahead of user demand, which continued to rise, and since it was really unsuited to full online timesharing, service continued in the batch processing mode. It was not until 1963, when the 7094 arrived to replace the 7090, that the Computation Center's research group, under the leadership of Dr Fernando J. Corbato, was able to offer full-fledged online timesharing as the standard service. In the timesharing mode, the 7094 operated under a set of Computation Center programs known as the Compatible Time Sharing System (CTSS), and could provide acceptable simultaneous service to 30 online users. So began a gradual switchover to the online operating mode, with remote user terminals connected to the central computer by means of telephone lines.

Meanwhile, so great was the demand for computer service at the time the 709 was being phased out, the Institute decided to purchase the machine rather than return it to IBM, and to set it up in a separate on-campus facility at a different location. The new facility was under the supervision of Dr John Slater, Professor of Physics, and Michael Barnett. The Slater facility operated on a fee-for-service basis, and for a while the CAD project used this installation. We were thus able to obtain much more computer time and shorter turnaround times than would have been possible at the Computation Center.

Another event that had an impact on the CAD project was the Institute's decision in 1958 to install on campus the TX-O computer. Originally developed at Lincoln Laboratory two years earlier by Wesley Clark and Kenneth Olsen, the TX-O was among the first computers built with transistors; it was

developed in 1956

*The director of the Computation Center was Dr Philip M. Morse, Professor of Physics. An excellent description of the Computation Center's origin and operation is included in Morse's book: *In at the Beginning*. See bibliography.

being replaced at Lincoln Lab by a much larger version, the TX-2. Once it had been determined by Lincoln Lab that it would be contractually possible to transfer this older, smaller machine to the campus on an extended loan basis, it was installed at the Research Laboratory of Electronics, where its installation and operation were supervised by staff members of our Lab. The presence of TX-O clearly had a salutary effect on the CAD project, especially, as we shall see, on that part devoted to the development of interactive workstations.

Still another event that had a major influence on our project occurred in 1963: the Institute's decision to undertake a major research and development effort whose broad objective was to investigate ways in which online, interactive timeshared computers could aid people in a variety of disciplines in their intellectual work. The plan was to get as many persons as possible

TABLE 5.1. Milestones in the History of Electronic Digital Computers and Digital Computation at MIT from 1947 to 1969

Date	Event
1947	Project Whirlwind started at Servomechanisms Laboratory under the leadership of Jay W. Forrester.
1950	Whirlwind I computer becomes operational.
1952	Professor Philip M. Morse initiates a research program in machine computation.
1955	Professor Morse establishes the Computation Center and becomes its first director.
1956	IBM offers a 704 computer to the Institute. One shift would be for MIT users, a second for other New England colleges, and the third shift for IBM.
1957	IBM 704 installed in the Compton Laboratory, Building 26.
1958	Whirlwind computer phased out for general computational purposes. The TX-O computer is installed on campus.
1960	IBM 709 replaces the 704 at the Computation Center. Research on online timesharing of a central computer begins under the leadership of Professors Herbert Teager and John McCarthy, and Dr Fernando Corbato.
1961	*Ad hoc* Study Committee on Computation issues its report on "Long-Range Computation at the Institute." Among the recommendations: a long-range research program on machine computation to make possible easier access to a central computer facility through use of remotely located terminals. In November, Dr Corbato demonstrates a prototype central timesharing system.
1962	IBM 709 replaced by the 7090 at the Computation Center. The 709 is purchased by the Institute for a separate computer facility under the supervision of Professor John Slater and Michael Barnett.
1963	Project MAC begins under the leadership of Dr Robert M. Fano, Professor of Electrical Engineering.
1964	IBM 7094 replaces the 7090 at the Computation Center and becomes operational as an online timesharing computer serving up to 30 online users simultaneously.
1968	Project MAC's GE 645 computer, operating under MULTICS, an advanced online timesharing software system, becomes operational. Intensive development work continues.
1969	MULTICS service is offered by the Information Processing Center for general MIT use. System capacity would increase during the 1970s to over 100 simultaneous users.

from all schools and departments of the Institute involved in the use of a computer, whether for research, development, management, or educational purposes. The expectation was that such widespread usage would lead inevitably to new applications and eventually to a new, advanced generation of computers (5–9). The project was sponsored by the Office of Naval Research acting on behalf of the Department of Defense's Advanced Research Projects Agency (ARPA). It was housed in its own quarters and was under the leadership of Dr Robert M. Fano, Professor of Electrical Engineering. The project became known as Project MAC, variously considered to stand for Machine-Aided Cognition, Multi-Access Computers, or Maximum Assembly of Components.* Not only did our Laboratory become a heavy user of the MAC computer (which started out as an IBM 7090 operating under the Computation Center's CTSS software), it also contributed heavily to the expansion of the 7090's peripheral facilities through the CAD Project's workstation development task. MAC's 7090 was later replaced by a 7094; still later, the CTSS software was replaced by a more advanced timesharing system called MULTICS, standing for Multiplexed Information and Computing Service. MULTICS was implemented by Project MAC on a General Electric 635 computer specially modified for MULTICS, designated, after modification, GE 645. Eventually MULTICS enabled more than one hundred users to be online at the same time. The events described in the preceding paragraphs are summarized in Table 5.1.

Interactive Workstations

In light of these interrelated events, we now turn to a second task of our CAD project: interactive workstations. As our early discussions with members of the Mechanical Engineering Department's Design Division continued, it became amply clear that some kind of interactive workstation would be needed for the computer-aided design environment. The designer would require an electronic substitute for his traditional pencil, sketchpad and drawingboard. Although each creative designer has his own method of working, there appears to be a common thread. The physical shape of the object being designed initially takes form in the designer's mind; this mental image of the part is then transferred to paper as a free-hand sketch, which serves to inspire refinements in the design. Successive iterations of this mental-image to paper-sketch process culminate in a trial design. To be minimally acceptable, an electronic workstation would have to support these thought-to-image interactions and allow the designer to feel comfortable in the new environment; it was hoped that a well-designed workstation backed up by a powerful computer would actually speed up this process. Incidentally, our term for the man–machine interface system was not workstation—that term surfaced many years later. We called it the ESL Display Console (ESL standing for Electronic Systems Laboratory), but it was more frequently referred to as the Kludge, presumably because of its complexity and the vari-

*Or, facetiously, "*M*ore *A*ssets for *C*ambridge."

ous contraptions it utilized to input, manipulate and output information pictorially.*

The basic workstation requirements finally agreed upon were these: it should be capable of accepting graphical and numerical data as inputs; it should be able to process the data either locally at the terminal or in the central computer to which it was connected; and it should be able to display the resulting processed data. From the human factor viewpoint, the terminal should be easy to use, respond rapidly to commands, and presumably not consume the entire capacity of the central timeshared computer to execute the designer's commands. It is not surprising, therefore, that these were the objectives set by Ivan Sutherland for the graphical communications system he developed at Lincoln Lab (in fulfillment of his doctoral thesis requirements (5–10)), which he implemented on TX-2. Since Sutherland had received extensive counseling from Coons on how designers carry out the design process, the views of Coons and his associates were strongly reflected in Sutherland's work. His graphics communication system was called Sketchpad, a landmark example of a man–machine interaction system.

Servo Lab's workstation development effort was under the leadership of John E. Ward, who also served for a while as project administrator for the entire CAD project. Ward was assisted by staff members and several graduate student assistants at the Laboratory. Coons' and Mann's graduate students conducted their own separate studies of the subject. Two general classes of problems had to be addressed: the specification and development of input and output equipment, and computer programs for data processing. Understandably, mechanical engineers and electrical engineers tended to look at these problems—initially, at any rate—in terms of their own disciplines. The two mechanical engineering graduate students, Johnston and Verderber, concentrated on electromechanical approaches for inputting and displaying graphical information, while Ward and his electrical engineering staff studied electronic methods for accomplishing the same end-objectives. Johnston produced an input technique in which the designer sketched projections of a mechanical part with a specially designed electromechanical drawing device augmented with a battery of pushbuttons that served to indicate numerical values. Sufficient orthographic information was thus conveyed to the computer to enable it to compute a complete shape description of the part (5–11). Verderber's work, conducted in parallel with Johnston's, involved a critical examination of three sketching systems, each requiring differing amounts of computer assistance. In one version, dubbed "the invisible drafting machine" by its originator, the designer specified the desired curve to be produced on paper (straight-line or circle) and its position with respect to a reference. The computer would activate a motor that caused a mechanical pencil device to sketch the curve. His second system was essen-

*There are many versions of what the word kludge means. Here is the author's definition, used in the context of the Laboratory's first display console:

Kludge is technical slang for a strange assemblage of a motley set of components that usually performs, to the amazement of all, in the way the designer intended it to.

tially an electronic tracing machine, the designer tracing over mechanical drawings of predesigned parts with a specially designed tool, resulting in a digitally encoded representation of the drawing. The third approach was, in some respects, an electronic version of the invisible drafting machine: the designer had three cathode-ray-tube screens; he could move three index spots, one on each display, by means of a joystick. The computer's role was to calculate and display these three self-consistent orthographic views of the part under design as the designer moved the index on one or another of the three screens. It was envisioned that the designer could record the index point on any one of the screens, could *connect* points to form a line or a circular or parabolic arc, could *erase*, could *specify dimensions* of lines, and could ask for dimensions between specific points for each orthogonal view. Clearly there was much sparring with the general man–machine interaction problem at this stage of the workstation development effort as the participants sought the "best way" to communicate information back and forth between the designer and the computer.

The presence of the TX-O on campus contributed in a major way to our thinking on interactive displays. It also changed the attitudes of the Institute-wide user community toward interactive computation. One reason for this change was the openness with which the facility was run, and the opportunity it offered users to interact personally with the computers. Students especially were given strong encouragement to use the machine, which was available around the clock; any *bona fide* student or staff member could sign up to use it for a maximum of one hour at a time. Thus, each user quickly developed a sense of having. a computer all to himself—in other words, having access to a *personal computer*. Thus, the environment of the TX-O contributed markedly to the pressures that were building among on-campus users for quicker turnaround computer service, and for service that would give them a sense of communicating directly with the machine.

Once students and others were given free rein on the TX-O, there emerged a whole spectrum of unusual programs for a wide variety of applications.* Ward and his group used the TX-O as an experimental vehicle in their development of improved light pens. Both Ward and Ross, through their Project Whirlwind and Lincoln Lab associations, were familiar with prior technology that employed a light-sensitive element or device as a sensor of images on the face of a cathode-ray tube. An early device at the Whirlwind facility was known as a light cannon; it consisted of a photomultiplier tube and associated electronics permanently mounted above the screen of a cathode-ray-

*Not all the programs were for research or academic purposes. Many students, overwhelmed by the potential of the machine, would spend many sleepless nights "hacking around" and writing programs that expressed their imaginative ideas. One such program, called Space War, was an interactive program in which two operators, each with a joystick and pushbutton, could manipulate an object on the screen (a space ship) and try to shoot down his opponent by releasing a rocket at the strategic moment. The Space War program bore a close resemblance to modern interactive video games. It actually was implemented on a commercial version of the TX-O, the PDP-1, donated to the Electrical Engineering Department by Digital Equipment Corporation in 1961 and installed adjacent to the TX-O.

tube console. Unwanted images displayed on the screen could be blanked out by covering them with an opaque material, thus having only the desired parts of the display for sensing by the light cannon and subsequent processing by the computer to which the cannon was connected. Whirlwind also employed a "light gun" in conjunction with its displays. The gun, which was rather bulky, looked like a snub-nosed toy pistol, complete with a trigger which was activated to expose the gun's photosensitive pick-up element to a CRT luminous spot. Ross and Ward had used the Whirlwind light gun in conjunction with their digital in-flight instrumentation project, mentioned in the preceding chapter.

In the mid-1950s, Lincoln Lab's TX-O computer group developed a light pen system that included a computer-controlled CRT display, a Photodiode sensor in a pen-like holder, a high-gain amplifier, and sampling circuits for digitizing the amplifier output (5–12). In fact, Lincoln Lab's TX-O computer included these features when it was transferred to the MIT campus in 1958. In Fig. 5.4, the light pen is being used by Earl W. Pughe, Jr, in the on-campus TX-O installation. Pughe, a member of our Laboratory staff, was the on-campus TX-O supervisor from 1958 to 1959.

Despite the availability of a light-pen design at Lincoln Laboratory, Ward and his group continued development work on the device, particularly to take advantage of the latest miniaturized components and to narrow the

FIG. 5.4. The on-campus TX-O Installation.

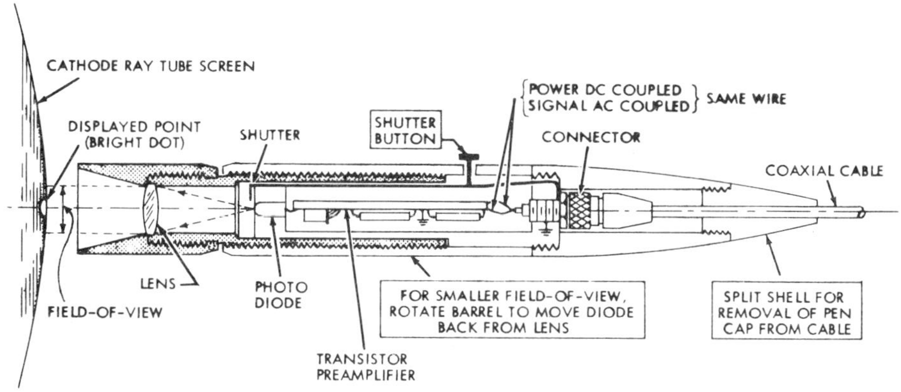

FIG. 5.5. Simplified mechanical layout of focusing light pen.

pen's field of view. The curvature of the face of cathode-ray tubes, and the need to cover the face with a protective transparent plate, often resulted in a separation distance of some two inches or so between the CRT face and the light pen, causing an intolerable amount of resolution loss. In a redesign, resolution was enhanced by inclusion of a focusing lens in front of the pen's light-sensitive diode—a rather delicate job because of the need to keep the pen slim and lightweight. The final design, illustrated in Fig. 5.5, was a pen approximately ⅝ inch in diameter and 7 inches long—not much bigger than a vintage-1940 fountain pen. It was part of both the TX-O display system and the ESL Display Console associated with the Project MAC computer. This focused-light pen attracted so much attention inside and outside the Institute that Ward agreed to build more than a dozen to satisfy requests for duplicates. Among those requesting and receiving pens were Lincoln Laboratory, the Brookhaven National Laboratory, and the Air Force's Cambridge Research Center, as well as several MIT groups.

The interactive graphics display console that finally met the requirements set forth by our CAD research group is shown in Fig. 5.6. Initially, in 1964 it was connected directly to Project MAC's 7094 timeshared computer by means of some fourteen data communication lines. Later, advanced versions became part of MAC's MULTICS timesharing system. Because of its versatility, this terminal—the Kludge—was the first practical solution at the Institute to the man–machine interface problem posed by creative computer-aided design requirements: it was a general purpose interactive facility that served the research interests of many diverse groups over its life span. A small sample of its applications included three-dimensional molecular model building; graphical plotting of numerical data; the design of highway intersections; electronic circuit design; ship hull design; modeling of the human vocal tract; and graphical portrayal of stress contours during the mechanical design process. Figure 5.7 shows a sampling of the drawings that could be displayed on its oscilloscope screen. So heavy was the demand for time at this terminal that a second one was built and made generally available to the MIT community. In the mid-1960s, usage of each terminal averaged 16 hours per day.

As indicated in Fig. 5.6, the heart of the display terminal was a CRT which served as the electronic substitute for drafting paper. The useful dimensions of the tube face were approximately 10"×10". Designers could draw linear or curvilinear shapes on the screen by means of the Laboratory-designed light pen (barely visible lying on the table in Fig. 5.6, but shown clearly in Fig. 5.8). With the aid of the control pushbuttons at the left side of the screen (see Fig. 5.6), the light pen was able to "drag" the CRT's luminous spot across the tube face, creating a straight or curved trace. Precise motions were unnecessary. Once end points were specified, the pushbuttons could be used to connect them with smooth contours—straight or curved lines.

The display terminal not only served as a man–machine interface for the many individuals interested in exploring online, interactive computers as aids to their research, it also proved a valuable experimental vehicle for our terminal-design group as they sought to discover preferred methods of communication between the operator and the computer. Feedback from individual users provided invaluable insights about the convenience of various types of data manipulators. Thus, in Fig. 5.6, two methods were available for controlling the movement of the objects on the screen: the hemispherical globe in front of the screen and the control box at the far left in the picture.

FIG. 5.6. The interactive graphics display station called the "Kludge."

The globe-like structure, which fitted nicely into a cupped hand, behaved as a joystick with three degrees of freedom: it had limited bi-directional rotation about three orthogonal axes—two horizontal and one vertical—and was programmed to control the rates and directions of screen-figure motion around these axes in synchronization with globe motion. The circular knobs on the control box (shown at the far left in the picture) were used to rotate objects on the screen up, down, right, left and to scale them up or down in size. The teletypewriter (far right) was a standard MAC terminal that served as a keyboard to input alphanumeric characters and to control graphics programs.

The electronics for manipulating the graphics described above and for providing an interface with a data channel of the central computer were contained in the three-rack cabinet shown behind the console in Fig. 5.6. Although this configuration added considerable electronics to the overall display system, it represented a substantial forward step in workstation design. Other interactive computers typically processed data for their graphical displays in the computer itself, thus tying up a substantial amount of machine capacity for this task. Ward and his group sought to transfer a major portion of the display's computational load for real-time interactions to special-

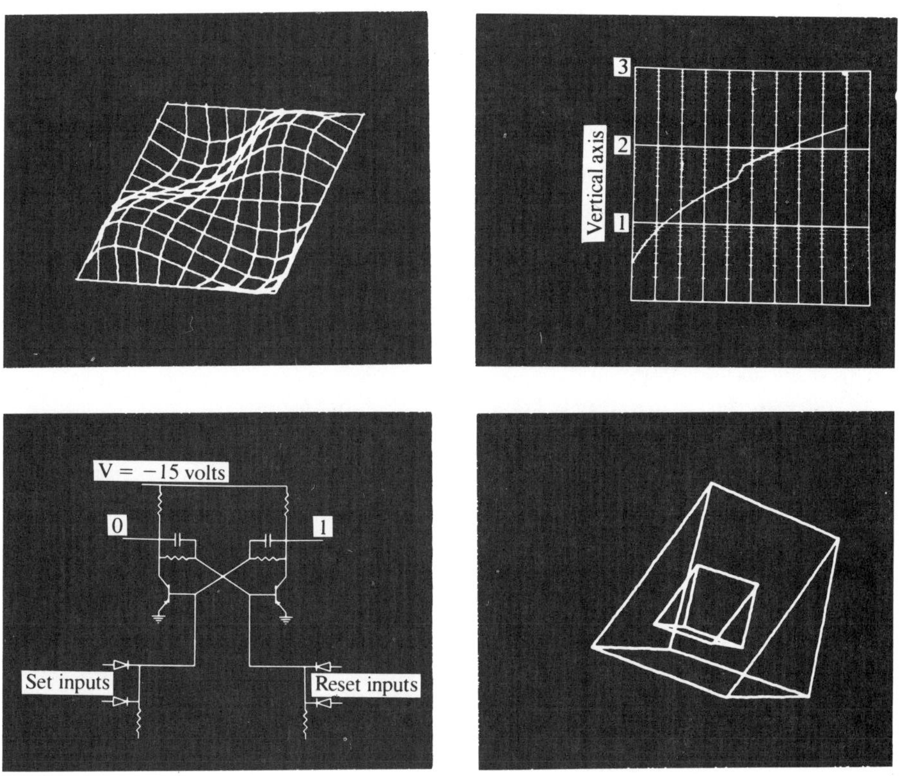

FIG. 5.7. Examples of graphics displayed on the Kludge.

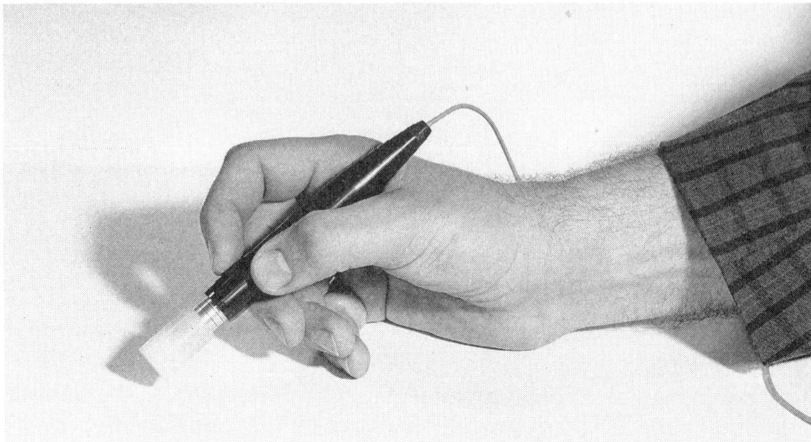

FIG. 5.8. The laboratory-designed focusing light pen.

ized circuits associated with the display while retaining supervisory control over the kinds of computations performed through programs stored in the central machine. The result was a display system that operated in a *semi-buffered* mode. This relegation of most of the computational details associated with graphics to local electronics released a substantial part of the central computer for other purposes.

The Automated Engineering Design Programming System

The Laboratory's graphics display work is temporarily set aside at this point to discuss the details of the CAD project's third component: the automated engineering design (AED) programming effort. Ross's experience in the development of the special purpose APT language for programming NC machine tools quickly led him to conclude, as the CAD project got under way, that existing programming systems were totally inadequate for the highly complex job of executing creative designs with computers as design aids. He reasoned that the myriad devices and systems which designers are called upon to synthesize and the highly unstructured nature of creative design called for a radical departure from the programming systems then in vogue. He envisioned a general problem-solving *system of systems* that would be composed of a number of systems programs organized in an internally coherent hierarchical structure. At the user end, the designer would be able to communicate and receive design information in familiar, simplistic terms. At the opposite end, the machine would be capable of executing highly complex operations, including automatic compilation of applications programs appropriate to the job at hand. In the middle of the hierarchical structure, there would exist applications programs in table-driven form which would be easily modifiable to suit a wide variety of situations within a problem class. In its ultimate embodiment, the entire programming system would be both problem- and machine-independent, meaning that a variety of design applications could be accommodated on a variety of computers.

While it is easy to state Ross's goals, achieving them was a different matter. It was first necessary to develop a wholly new philosophy—a global view—of a programming system that would enable designers to cope with any conceivable design problem, many of which would call for widely disparate procedures. Such a philosophy could not be whimsical; it must be soundly based for it would be the bedrock upon which problem solving in the computer environment would be built. Having formulated and adopted a broad, generalized philosophy for his programming system, Ross's next challenge was its implementation: a formidable task indeed, since his global philosophy led him to conclude that a completely new and nontraditional programming language would have to be developed. As it turned out, many tens of man-years of programming effort would be expended before a practical finished product was available. Seven calendar years were to elapse before there evolved, as Ross expressed it: ". . . [a] system of [programming] systems for making systems, along with an orderly method for applying them." Ross's programming philosophy, and his collection of concepts and working tools, became known collectively as the automated engineering design (AED) approach.*

Ross's concept of a generalized programming language system emerged slowly. It began to take shape in 1960, as the APT project was gradually phasing out. A key influence on the system came from the redesign of the arithmetic element (ARELEM) of the APT system, undertaken at the request of the Aircraft Industries Association between 1960 and 1962. The radically new data structures and processing techniques employed in the ARELEM redesign became part of the AED system approach. The new version of ARELEM allowed completely arbitrary cutter shapes whose profiles approximated arbitrary concatenations of straight-line and circular-arc segments, and permitted use of any number of arbitrary geometric surfaces to influence the determination of cutter motions. Observe that the notion of *generalized* concepts figured prominently in the redesigned ARELEM; however, since APT was a *specialized* language, tailored to a specific task (programming machine-tool motions), it was much too narrowly oriented for use as a general problem-solving tool. A still-more-general programming system had to be devised that would accommodate APT and many other specialized systems, general within themselves.

What really lay behind the system of programming systems Ross ultimately devised was the development of a wholly new computer language, a job whose magnitude can only be compared to that of creating one of our existing spoken languages, or a new language—Esperanto, for example. In his paper (5–13), presented at the 1967 national conference of the Association for Computing Machinery (ACM), Ross described the four principal

*Since the language first chosen for development work and experimentation was a modified form of ALGOL-60, AED was sometimes interpreted to mean ALGOL Extended for Design.

phases involved in his implementation—the lexical, parsing, modeling, and analysis phases—as well as features of the processors required for each phase. Readers who are interested in the many details involved in these phases, as well as in the AED approach as a whole, are referred to this paper. We limit ourselves here to a short discussion of the highlights in order to convey the magnitude of the development that was undertaken.

First, with respect to global language philosophy, Ross postulated that in order that such a language be suitable for solving a problem, there must exist (l) data, (2) a structure to the data, and (3) a set of interpretive rules, called *algorithms*, that make the data and its structure meaningful. The rules should allow both the structure and the data within the structure to be manipulated and should permit the results to be related to other data structures. From this viewpoint, all computation, data processing and problem formulation must be considered in terms of data, data structures and interpretive algorithmic rules. It was within this framework that the AED family of programming systems emerged.

With a global philosophy about a general problem-solving system established, the next step was to postulate a substructure that not only reflected that philosophy but also allowed creation, modification and maintenance of that substructure with relative ease and efficiency. As already stated, Ross identified four distinct phases for his substructure: the lexical, parsing, modeling, and analysis phases, shown in Fig. 5.9. In the lexical phase, input statements pertaining to a problem are broken into discrete items that constitute the smallest indivisible elements of such statements—the analog of English *words*. The parsing phase enables the lexical elements to be grouped into phrases and sentences in order to determine the structure of the statements. The modeling phase extracts meaning from the statements and formulates an understanding of the problem posed by the statements. The analysis phase carries out the solution to the problem as interpreted or understood. A key feature of the AED approach was that each of the four phases was executed by means of its own processor, which was *table-driven*, meaning that each processor was governed by a list of controls (instructions) whose tabular content could be set up to fit the specific needs of a given application system. Thus, a variety of application systems could be constructed, one at a time, simply by appropriately modifying the table.

Developing control information capable of handling such complex, rich application languages proved to be an extremely complicated task. As Ross

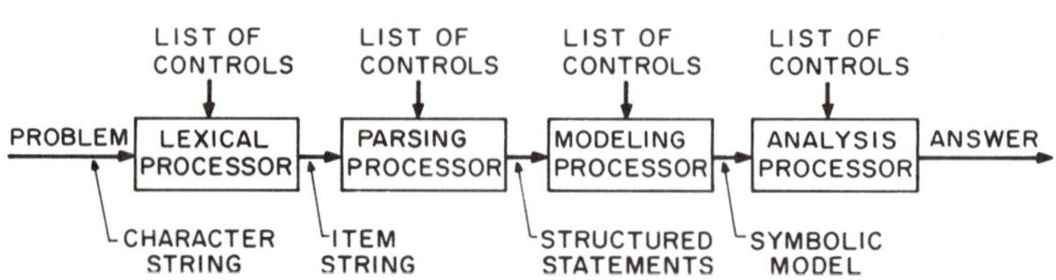

FIG. 5.9. The four phases of Ross's general problem-solving process. The lists of controls depend on the design application.

expressed it: "We may think of the generalized processor for a given phase (and there were four of them) as a strange kind of computing machine, and the setting of control information as writing a machine-language program for that strange device." To lighten the programming burden, the project group first set about developing a *compiler*, a powerful programming tool that would automate the assembly of the program information contained in each processor into machine-language code. Since the group had no previous experience with compiler development, its first effort in that direction was really only for educational purposes. However, that effort yielded such valuable insights into compiler development within the framework of the generalized problem-solving concept that a reworking of it produced a highly useful compiler system, designated AED-0. This system, however, was machine-dependent; it would run on Project MAC's timesharing machine but it was not compatible with normal IBM environments. Although originally intended for use only by the CAD project group for developing other programming systems within the AED family, a reworked version proved so powerful and practical that eventually it was made available to user groups both inside and outside the Institute.

One such group in Servo Lab, which was working on computer-aided design of electrical and electronic networks under different sponsorship, found AED-0 to be ideally suited to its programming needs and made extensive use of it. A few words of explanation are in order here about the origin of this second CAD project, under the leadership of Professors M. L. Dertouzos and J. F. Reintjes. As we watched the Air Force sponsored CAD project unfold, it was clearly evident that timesharing digital machines equipped with graphical workstations would be powerful design instruments in the electrical as well as the mechanical domain. We concluded, however, that in order to explore machine-aided electrical and electronic circuit design, separate research sponsorship would have to be procured, since our Air Force sponsor was primarily interested in CAD principles directed toward mechanical devices. Although there was some mild interest in the electrical area too, it was apparent that very little, if any, financial support would be forthcoming from the sponsor for this line of research. We therefore put together a separate proposal (5–9) on computer-aided design of electrical networks, a project that was eventually funded by the National Aeronautics and Space Administration (NASA). Our first grant was for $40,000, clearly far too little to make any observable dent in such a complex problem but entirely adequate in light of the supplementary support we could get from the Project MAC facilities and the Air Force's AED development work. Thus, in 1964, there were two CAD projects underway in the Laboratory, one directed toward mechanical design and the other toward electrical-circuit design.

The NASA-sponsored project had several goals, similar in many ways to those of the mechanically oriented effort: we were seeking the most effective way to partner an electrical circuit designer and a digital computer in an on-line man–machine interactive system, and we also wanted to develop an application programming system for electrical networks which would be feasible—and expandable—for designing increasingly complex linear and nonlinear networks. (Note the parallelism between the generalized programming system goals of the electrical- and mechanical-design projects.) We also wanted to explore network analysis techniques in light of physical constraints imposed jointly by the network configuration itself and the capabilities of timeshared computers. In the back of our minds, we hoped to accelerate the electrical circuit design process by enhancing the confidence level in trial designs at the computer-aided analysis stage. As a result, one might expect to minimize the amount of trimming and fine-tuning required at the so-called breadboard (unpackaged) state or even eliminate breadboarding altogether.

By 1965, two electrical network application programming systems had been developed, both written in AED: AEDNET and CIRCAL-1. AEDNET, developed under the leadership of Dr Jacob Katzenelson (now Professor of Electrical Engineering at the Technion-Haifa Institute of Technology, Israel), centered on the concept of naming a *device*, which would be composed of one or more circuit elements and have a specific configuration and input–output characteristics. Once the device was named, and given a special name

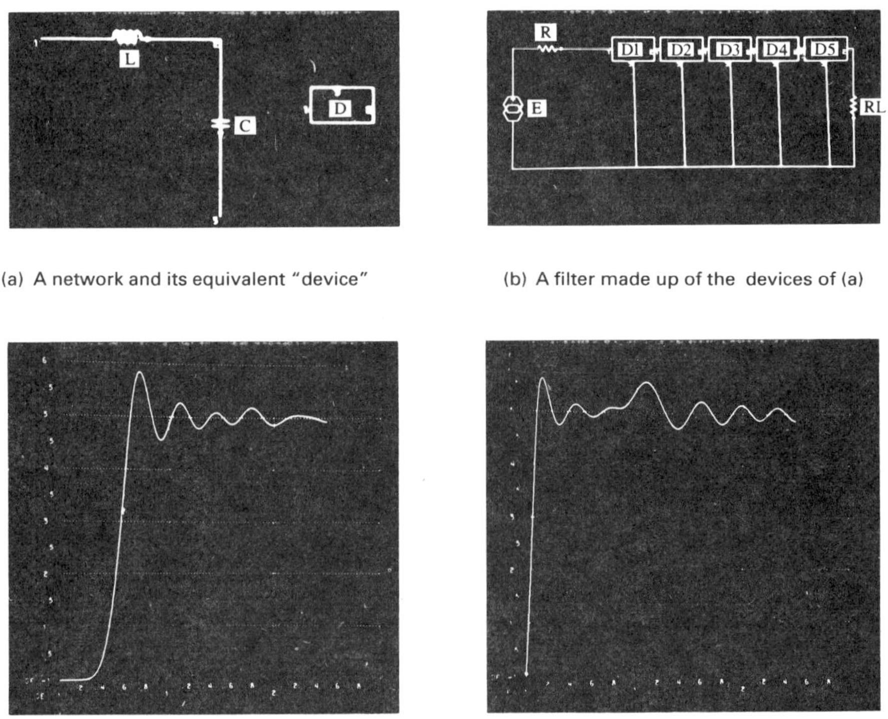

(a) A network and its equivalent "device"

(b) A filter made up of the devices of (a)

(c) Output voltage in response to a step input

(d) Input current

FIG. 5.10. AEDNET permits a subnetwork, such as a filter, to be defined as a device (a); several such devices are used to build a large filter (b); performance curves (c, d) are produced in response to a step-voltage input.

and symbol, it could be combined with other devices and used to construct larger networks. Both linear and nonlinear circuit elements could be modeled this way. Figure 5.10 illustrates the synthesis process. In Fig. 5.10a, inductor L and capacitor C are connected in an electrical filter configuration. This filter is then connected to similar filters to make a larger filter, as shown in Fig. 5.10b. Through the use of an analysis program, it is now possible to display on the workstation screen the output voltage of the filter when a sudden constant input voltage (a step voltage) is applied at the input end. All parts of Fig. 5.10 are photographs of the workstation's display screen (5–14).

The second CAD group, under the leadership of Professor Dertouzos, took a different design approach. In an early version of its work, called CIRCAL-1, standing for Circuit Calculation, a circuit was synthesized on the workstation screen through use of the light pen and a set of pushbuttons that bore labels designating basic active and passive electrical circuit elements—resistors, inductors, capacitors, diodes, independent and dependent voltage and current sources, and arbitrarily designated elements with specialized input-output behavior. To bring up an element on the CRT display, its location was designated with the light pen and its nature specified by pushing the appropriate button. The element's value was then inserted into the computer from the teletypewriter station. Through repetition of this procedure, complex circuits could be built up (see the example in Fig. 5.11).

FIG. 5.11. Professor Dertouzos (right) and Mr Therrien demonstrate their computer program for analyzing and displaying electrical-network configurations.

Upon completion of this network assembly phase, the analysis phase was initiated and executed; following this process, the designer could request a plot of the voltage between any pair of nodes in the network (5–15).

CIRCAL-l's uniqueness resided in the method used to compute network behavior. Rather than describing performance by a set of network equations, which were then solved by the computer, a computational model of the network was synthesized within the computer in the following manner: for each set of network elements, a computational model (data structure) was established and assigned a block of storage; the AED-pointer technique was then used to interconnect the stored blocks in direct correspondence with circuit connections. With a network model established, a state-space approach was used to calculate the network's dynamic behavior: that is, time was divided into small increments and the state of the network output variable calculated for each increment, with the network state at the end of each preceding increment taken into account. Electrical network design along CIRCAL-1 lines was sufficiently promising by 1965 to warrant continuance of the research in order to enhance the power of the technique. The folow-on research culminated in CIRCAL-2, an advanced version of CIRCAL-1.

Meanwhile, the experience gained by Ross and his group in designing AED-0, as well as inputs obtained from AED-0 users, provided a strong base upon which to build an upgraded version. This elaboration became known as AED-1, a truly workable general problem-solving system. However, in moving into AED-1 research, the group faced serious financial and manpower obstacles. In order to achieve a practicable general problem-solving system, AED-0 had to be reworked, and many features had to be added in the process. Although our sponsor continued to provide financial support, it appeared that such support would be forthcoming only in an amount to sustain the existing level of effort, not the factor of 10 or so increase which was needed to show tangible results within two or three years. Furthermore, it was unlikely that we could find and hire the required number of highly skilled programmers. It was at this juncture, in December of 1963, that Ross once again turned to the plan he had employed so successfully to complete the APT development: an invitation was extended to a group of industrial organizations in the United States and abroad to participate with us in a cooperative AED development effort. If received favorably, the plan would accomplish two goals: it would greatly speed up our work and, at the same time, satisfy our ambition (and that of the sponsor) to get industry involved in the work. It was hoped that, through this cooperative arrangement, our research results would flow smoothly from the Laboratory to the industrial sector. The invitation stressed the need for a nominal one-year commitment of an experienced system programmer from each company; this person would receive an appointment as an MIT Visiting Staff Member and would be the guest of the Electronic Systems Laboratory and Project MAC. Once more industry demonstrated its interest in joining forces with us. By March of 1964, six US companies—Boeing, Grumman, North American, Lockheed, Sandia, and United Aircraft—had nominated experienced system programmers for a one-year tour of duty with us; a seventh member, from IBM, arrived a few months later. This core group, together with our own staff, constituted the Cooperative AED-1 Project Group.

Although the original intent had been to start research immediately on an entirely new AED-1, using AED-0 only as the programming vehicle, there was a delay. It soon became evident that if we were to make good on our offer to the participating companies to have AED-0 available to them as a batch-processing system for the IBM 709, 7090 and 7094 computers, more enhancement would be needed than we had originally anticipated in order to make it a useful tool. This improvement effort required a full year, but in March of 1965, a greatly expanded AED-0 system was ready for release to interested companies, "embodying the most powerful high-level programming system available," as Ross described it. The enhanced version was IBM-compatible and ran in both the timesharing and batch modes (5–16); it was the first system and language that could be used for what is now called *software engineering*.

Enhancement of AED-0 was only part of the project effort during the 1963-65 time period. A system configuration for the still-more-powerful AED-1 was blocked out, and programming for some of the major blocks was completed. AEDJR, a working version of an applications-oriented programming system, was also completed and externally distributed. AEDJR was a general system that allowed specialized problem-oriented programming languages to be defined and put to use.

The Years 1965-1970

The Department of Mechanical Engineering's Design Division withdrew from the project in 1965, leaving Servo Lab's two active efforts—interactive graphical-display research and continued development of the AED language—as the focus for the final five years, both pursued in collaboration with Project MAC as co-sponsor. We will discuss first the display research, then AED.

Interactive Graphical Displays

Although the Laboratory's semi-buffered display console was very successful, several problems surfaced as changes were made in the central computer to accommodate increased demands for service. For example, the installation of an auxiliary high-speed drum memory resulted in blockage of Kludge's data whenever signals were swapped between drum and main core memory, resulting in an annoying blinking of the CRT screen every few seconds. Also, as the user load increased, the heavy demand for supervisory space in main memory left an inadequate amount for storing the display control programs and picture data. Furthermore, it was now envisioned that workstations in commercial environments would most likely be remote from the central computer, a configuration that complicated matters because of the need for relatively long, noise-free, low-distortion, high-speed communication channels between the central computer and the terminals.

Influencing the group's search for a solution to Kludge's memory problems was the fact that, in the 1960s, computer memory came at a very high price.* Memory was such a precious commodity that computer engineers avidly pursued every avenue that might possibly reduce the amount required for "overhead" operations in the timesharing environment. Unfortunately, Kludge consumed both supervisory memory and supervisory running time; an ensemble of data and programs had to reside in supervisory memory for refreshing the CRT display, coordinating signal flow between the display and main memory, screen editing, generating messages, and manipulating graphical data. In the early version of Project MAC's central timesharing system, a single Kludge consumed an overhead of about 10 percent of the central machine's supervisory memory and from 3 to 10 percent of the running time, depending on the amount of display.interaction. In addition, part of a second memory bank was required to handle Kludge's users' specialized programs, which had to be cycled in and out of secondary memory in accordance with the timesharing doctrine.

In light of these issues, there gradually emerged a consensus that a buffer computer, in the form of a small general purpose machine, should be inserted between each display terminal and the central machine so that both the display-data storage and Kludge's supervisory functions, presently being performed by the central machine, could be transferred to the buffer computer. The practicality of this approach was reinforced by the appearance, in the mid-1960s, of powerful small computers in the $20,000 to $50,000 price range. Hence, in late 1965, the Laboratory's display group recommended that Project MAC procure such a machine for experimental buffering purposes. The machine chosen was Digital Equipment Corporation's PDP-7. The choice was an easy one, since the display console itself had been constructed with Digital's logic modules; with a DEC machine, compatibility problems would be minimal. The fully buffered display system went into operation in August of 1967.

In the configuration that finally evolved, virtually all the required overhead operations were relegated to the buffer, thereby substantially lessening the computation burden on the main computer (5–17). What we had, therefore, was what today is commonly called a "smart terminal"—a small general purpose computer and CRT display console serving as the terminal of a large mainframe computer.† Although putting a small computer between

*Making direct comparisons between memory costs in the 1960s and present day costs is complicated by the vast: changes that have occurred in the technology over the years an our election in later years to use the byte (equal to eight bits), rather than the bit, as a reference unit. In terms of raw numbers, however, the cost per bit of secondary (disk) memory is now roughly 1/100th of what it was twenty years or so ago for comparable capacity drives. For the, maximum-capacity drives available at each time period, the reduction factor is approximately 1/1000.

†The companion term "dumb terminal" refers to a terminal with no buffering or processing capability between the terminal and the mainframe. The "smart" and "dumb" descriptions crept into computer-buff jargon some time after the conclusion of our project.

each Kludge display console and the central timesharing machine solved the response-time problem, a different problem popped up—the improvement in response time came at the price of having an extra computer for each console. There began, therefore, a long-term effort which endured for many years, both inside and outside MIT, to achieve interactive graphical displays of sufficient versatility and acceptably low cost to make them the preferred input–output devices for both timesharing and stand-alone purposes. In retrospect, Kludge development served to highlight the enormous power of the real-time graphical display as a problem-solving tool in a spectrum of disciplines, as well as the urgent need for research that would bring its cost into line with the conventional inputting device of the 1960s, the teletypewriter. Fortunately, the dramatic technical progress made over the past 25 years in the fabrication of large-scale and very large-scale integrated circuits, and the steady reduction in their cost, have finally produced inexpensive graphical workstations.

The research conducted by Ward and Robert H. Stotz on low-cost graphical displays was an early first step in that direction. Realizing that memory and signal processing costs were major bottlenecks, they sought a way to minimize both. Their estimate was that memory alone for a refreshed graphical display was amounting to some $10,000. They concluded this figure

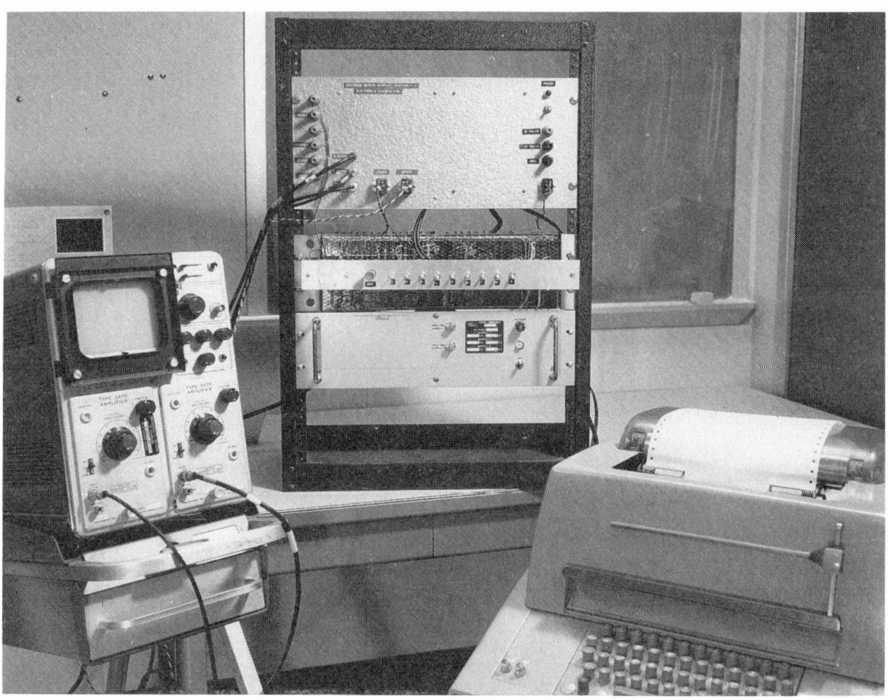

FIG. 5.12. ARDS-I.

would have to be reduced substantially if CRT displays were ever to become popular.

The tack taken by Ward, Stotz, and graduate student Tom Cheek, a third member of the Laboratory's display group, was to investigate CRTs that could display images on their screens without the need for cyclic refreshment—in other words, they began to look at the possibility of image-storage tubes. With storage tubes, an image is "painted" only once on the tube's face; thereafter, it is retained indefinitely through the action of a stream of

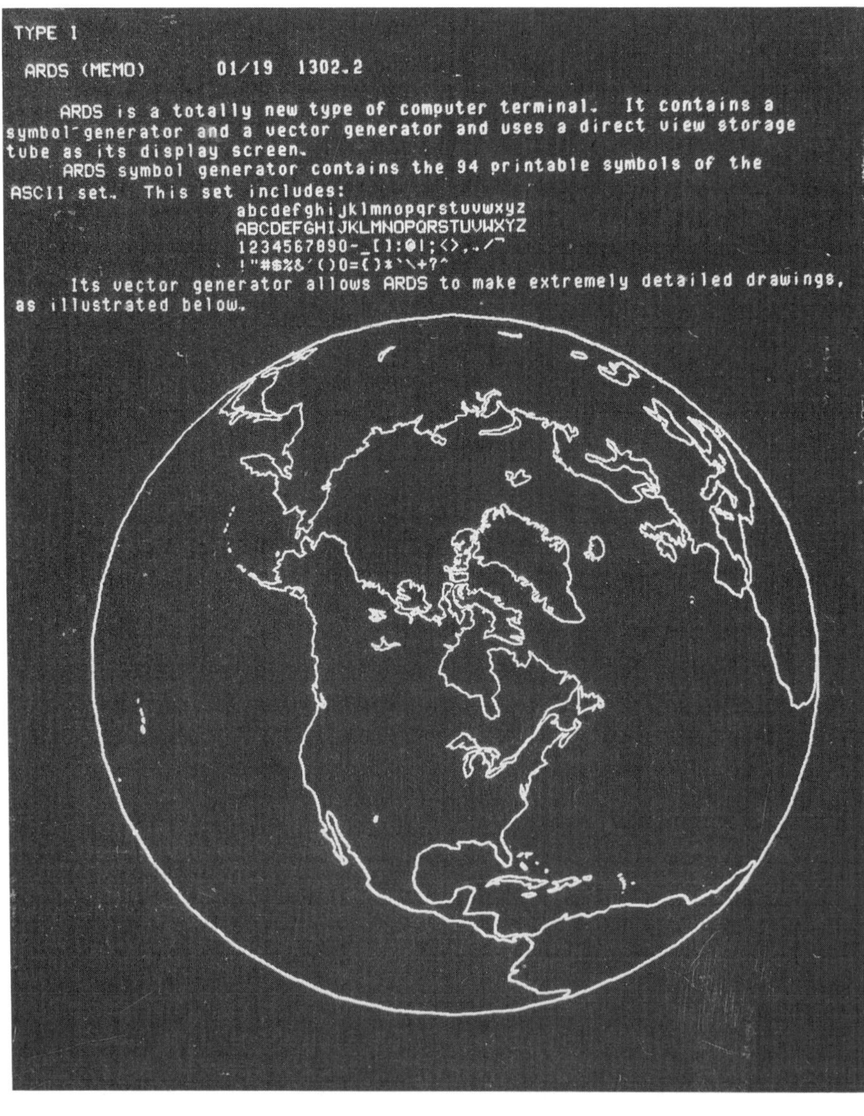

FIG. 5.13. An image displayed on ARDS-II.

"holding" electrons which continually floods the screen until the image is erased through action of an electron "flood gun." Thus, instead of an image having to be continually refreshed (rewritten) in order to be viewed, at a typical rate of 30 refreshments per second, the storage-tube image is written only once, thereby eliminating the continuous data access needed for screen refreshment. In computer parlance, the storage tube has built-in memory—it is, itself, capable of storing information. The display group characterized a storage tube graphical display as an electronic Etch-a-Sketch process.

Storage-tube technology was well known in the 1960s; it had been around since World War II, when such tubes were used to store radar-generated signals. In fact, as it existed in the 1960s, storage-tube technology was still quite adequate for radar applications and for image storage types of oscilloscopes, but only marginally adequate for computer display applications because the screens were too small and picture resolution was too low. Therefore, the immediate job was to acquire some specially fabricated tubes of a size which would not be too much of a letdown for Kludge users, who were used to a 10-inch screen. To demonstrate the practicability of this technology, and also, it was hoped, to excite the storage-tube manufacturer's interest in developing larger ones, the group assembled a display terminal around a Type 564 5-inch storage oscilloscope marketed by Tektronix Company. On this display, characters could be drawn at the rate of 120 per second and lines written at a rate of 30 to 60 per second (about ten times faster than teletypewriter rates). The display electronics were connected to Project MAC's central timesharing computer over a 1200-baud phone line, and inputting commands were issued from a Model 37 teletypewriter using a separate 150-baud phone line.* This experimental terminal, dubbed ARDS-I (standing for advanced remote display station and shown in Fig. 5.12), was sufficiently promising to warrant construction of complete terminal electronics to accommodate a 10-inch tube known to be under development by Tektronix. Six months were to elapse before this tube arrived, however. Meanwhile, Stotz and Cheek described the complete system in a conference paper presented at the eighth national symposium of the Society for Information Displays (5–18). The interest exhibited by the attendees at this meeting may have hastened the arrival of the 10-inch tube because, shortly afterward, Ward and Stotz were able to persuade Tektronix to lend us their 10-inch production prototype, which arrived three months before any 10-inch tubes began moving off Tektronix production lines.

The ARDS-II had a useful display surface of 6½"×8½" and a markedly improved resolution over the Type 564 unit. Up to four thousand 7×9 dot-matrix characters could be displayed. A typical picture display is shown in Fig. 5.13, and a commercially packaged version, complete with keyboard and auxiliary input device—the mouse—is illustrated in Fig.5.14. The

*In telegraphic transmissions, transmission rate is defined in terms of words per minute, there being 48 bauds per word. If it is assumed 48 bits are required, on the average, to digitally encode an English word, one baud equals one bit.

mouse, a Stanford Research Institute development, served to steer a non-stored cursor over the tube's face in accordance with horizontal and vertical positioning commands set up through activation of pushbuttons located on the mouse. ARDS was seen as a good response to the need for low-cost graphical displays, and displays of its type enjoyed a 12- to 15-year product life span—the time required by integrated-circuit technologists to reduce memory costs to the point where refreshed-screen raster displays could once again be price-competitive. However, during their product lifetime, storage-tube displays were enthusiastically received. Their crisp images were pleasing; resolution was excellent and pictures were solid and motionless compared to refreshed raster displays which tended to shimmer and reveal individual raster lines. The fact that refreshed displays could once again supplant their storage-tube counterparts and take over the market is evidence of an inherent weakness of the storage-tube approach: it was unable to respond quickly to small editorial changes which could be easily made with the refreshed-screen display. Although information could always be added to the storage tube display, any deletion, however minor, required that the entire picture be erased and rewritten, thus making the dynamics of the system annoyingly slow. In refreshed displays, this dynamic sluggishness is absent since the entire picture is automatically rewritten once every 1/30th of a second.

FIG. 5.14. A commercially packaged ARDS-II.

User response to the ARDS terminal was so favorable that Stotz and Cheek left the Laboratory in March of 1968 to found their own company, Computer Displays, Inc. (CDI), with ARDS-type terminals as their principal product. After approximately one hundred had been sold, CDI merged with ADAGE corporation, which marketed many more ARDS terminals. Part of the impetus for creating CDI had come from the same kind of situation we had faced sixteen years earlier when industry was reluctant to take on numerically controlled machine tool systems as a standard product line. The Laboratory display group was unable to interest any industrial organization in picking up on ARDS, as the industrial sector saw no market potential in storage-tube display terminals. Even Tektronix, a major manufacturer of the principal component of such a display—the storage tube itself—preferred to confine itself to supplying tubes to display system manufacturers and to refrain from entering the system business. As it turned out, Tektronix later changed its mind; sensing an increasing interest in storage-tube workstations, they first expanded development work on larger tubes and eventually announced their entrance into the storage-tube display system business, in competition with ADAGE. Tektronix, as the major supplier of the principal terminal component, the storage-tube, eventually came to dominate the market.

The interactive graphics display project provided excellent opportunities for student participation through graduate and undergraduate thesis research. Between 1965 and 1969, a total of 14 theses were performed on the project, ten at the graduate (MS) level and four at the undergraduate level. Of these theses, 12 were supervised by Ward and two by Ross.

AED-Language Research

While graphical display station research was proceeding under the combined leadership of Ward, Stotz and Cheek, Ross and his group were moving their AED language work forward with the assistance of visiting staff from industry. This cooperative plan proved so satisfying to all involved that, at the end of its first year, it had become an integral part of the AED programming effort; over a span of six years, a total of 31 programmers from 21 organizations here and abroad supplemented the Lab's permanent staff. The names and affiliations of these contributors are shown in Fig. 5.15.

It will be recalled that a major project goal was to turn the largely experimental AED-0 general purpose language system into a practical professional system that could be used with a variety of computer systems for general problem-solving. However, as the group moved toward that goal, it became evident that what was needed to do the job was not a single programming system but a "system of programming systems." Hence, as the research unfolded, the more elaborate this system of systems became, the longer it took to reach an end-product that would be machine-independent. By 1968, several systems had emerged: AED-1, RWORD, AEDJR, and CADET. The domain of AED-1 was general programming, and compiling and operating other programs; essentially it could be used on any large-scale computer,

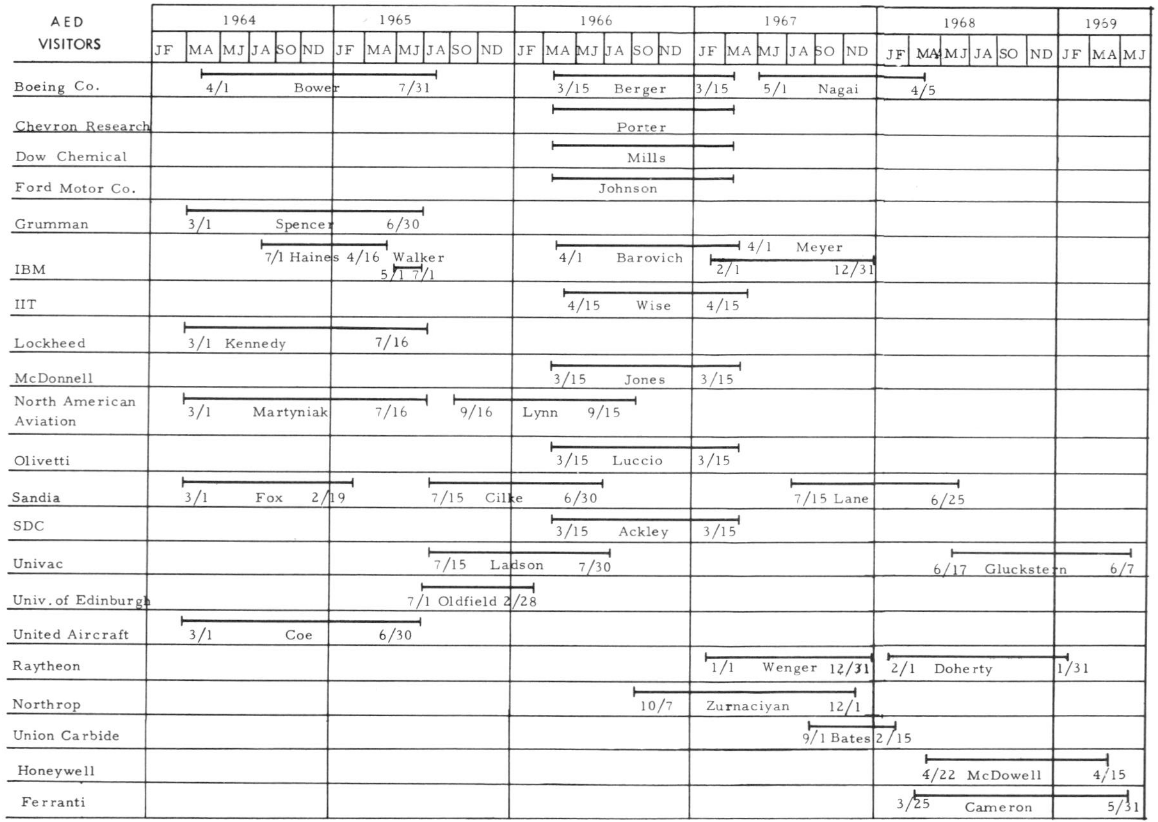

FIG. 5–15. AED visiting staff.

such as the IBM 7094, the Univac 1108 or the IBM System 360. RWORD's function was to build a lexical processor which would convert input in the form of computer-word strings to a form required by the AED system. As mentioned earlier, AEDJR was a sentence-parsing processor. The CADET system, standing for computer-aided design experimental translator, was a programming system aimed at those exploring generalized approaches to specific computer-aided design applications—mechanical, electrical, or whatever.

In keeping with the group's desire to make the results of its work available to the industrial community, various versions of the AED family of programs were released from time to time, beginning in 1964. Some 27 airframe, computer, government, and university organizations received these preliminary releases and some 16 received the final version (version 3) which ran on the IBM systems OS/360 and CP/67. In a further technology transfer effort, three conferences devoted exclusively to AED programming philosophy were held at the Institute: the first was presented on June 22–23, 1966, with 105 participants, the second was held on January 25–27, 1967, with 349 participants from 105 organizations attending, and the third took place on July 15, 1969 for 106 participants.

Many of the accomplishments of the CAD project team came together in July, 1968, on the occasion of a computer conference held at the Technical University of West Berlin (TUB), under the joint sponsorship of TUB, the Ford Foundation, Project MAC, and Servo Lab. For this meeting, we set up a communication link between a lecture hall at TUB and Project MAC's timesharing computer in Cambridge. Two communication paths were established: a two-way, 1200 bit-per-second link via Frankfort and New York for an ARDS terminal, and a 200 bit-per-second link via Paris and London for a teletypewriter terminal. During the meeting, Professor Dertouzos used an ARDS terminal to demonstrate his CIRCAL-2 electronic circuit design program which had been programmed in the AED-0 language. Needless to state, this confluence of three major elements of our project in a single demonstration brought us no little measure of satisfaction and delight; not only were we able to show tangible evidence of our research results, our advanced remote display terminal performed admirably in a truly remote setting—four thousand miles from its host computer! Figure 5.16 shows ARDS set up in a classroom at TUB, together with diagrams of the communication paths employed between West Berlin and Cambridge.

The CAD project came to a close in January, 1970, through a series of interrelated matters. Undoubtedly, a contributing factor was that two strong proponents of our NC/APT/CAD effort of the past twenty years, the Air Force's Webster and his colleague Max A. Guenther, were approaching retirement and no comparable replacements were waiting in the wings to assume their roles. More importantly, however, MIT had amply demonstrated the technical feasibility of computer-aided design and had provided, through the AED system of programs, a structure upon which others could build computer-aided design systems of their own. In keeping with the Air Force's

FIG. 5.16. The ARDS display station in the Technical University of Berlin. Diagrammed on the blackboard are the communication channels that linked ARDS to the Project MAC computer in Cambridge, MA.

policy to encourage further developments and refinements of the system within the private sector, once proof of feasibility was established, it was thought unlikely that additional financial support would be forthcoming from the Air Materiel Command beyond 1969. Webster and Guenther, while well aware of this policy, also knew that, like APT, the AED programming system would have to be maintained and continually upgraded if it were to survive. To ensure this, they decided to persuade the AIA to pick up the maintenance and upgrading effort, just as this organization had done so successfully in the case of APT. Thus, in February of 1967, Reintjes, Webster and Guenther traveled to San Diego to meet with Burt Raynes, President of Rohr Aircraft and an influential member of the AIA, to pursue that strategy. This time, however, we were not successful; after careful deliberation, the AIA responded negatively.

Fortunately, the AED system of programming survived despite the AIA's rejection because still another factor entered the picture. The entrepreneurial spirit—which so often awakens in the minds of creative people—came suddenly to life in the mind of Ross. So deep was his involvement in both APT and AED, and so convinced was he of the continuing need and opportunities for high-level programming language systems for general problem-solving, that he made the important decision to withdraw from MIT to form his own software company and to build upon his AED work as a member of the private sector. Joining him were several highly qualified and experienced members of his staff and one of his enthusiastic supporters from United Aircraft who had been one of the early participants in the industry cooperative program. Together they formed SofTech, Inc., a software company that still plays an important role in the development of complex programming systems.

Personnel and Contractual Matters

Over the ten years and a month from December 1, 1959 through January 31, 1970, a total of 146 people participated in the CAD project, some for the entire ten years, others for shorter periods. Included in this number was a total of 115 MIT faculty members, research staff and students and 31 industry representatives, each on campus for approximately one year as members of the AED cooperative program. In addition, another 560 persons from industry, the government, and universities attended the three technical meetings on AED in 1966, 1967 and 1969.

Over the same time span, MIT's contracts from the Air Materiel Command of the US Air Force amounted to a total of $3,108,400, or about $305,000 per annum. Although this annual figure may seem small relative to what was achieved by the project, including all the equipment that was developed and built, it was quite realistic when account is taken of the considerable additional support received from Project MAC in later years, both in staffing and use of its computer facilities, and the nearly 31 man-years of effort contributed by industry through the cooperative programs—a most valued adjunct to research staff resources.

6

THE ROLES OF THE AIR FORCE, THE SERVOMECHANISMS LABORATORY AND INDUSTRY

The numerical control project provides valuable insights into what it takes to create a new technology and establish it in the marketplace when at the outset no single organization is prepared to undertake the whole job. Even though segments of the machine tool industry had dabbled in electronic control for some years, no noteworthy product had emerged from those efforts at the time the Parsons Corporation sought financial support for its ideas from the Air Force. Even then, Parsons Corporation was unable to undertake the work, exclusively, because of the narrow spectrum of its internal skills. The academic research laboratory partially filled this void by providing the necessary intellectual insights to the problem, finding a technical solution by drawing expertise from its technical environment, and providing the programs required to transfer the new technology from laboratory to industry. Once this was accomplished, industry put numerical control on a firm base from which there emerged a vast array of digitally controlled processes and products in the marketplace.

Even so, it took more to ensure success than three entities—the government, an educational institution and industry—acting independently; it was the interaction among the three that yielded the successful formula. We must emphasize that numerical control was neither preplanned nor accidental: it evolved gradually. In retrospect we see that a series of crucial events and opportunities occurred along the way, and at each occurrence, the decision that was made contributed positively to the outcome. In this chapter we give details of the respective roles played in the course of the project by each entity: the Air Force, Servo Lab and industry.

The Role of the Air Force

As a prelude to our discussion of the Air Force's role in numerical control—and why it was interested in this technology—we need to look briefly at the world political condition in 1949 and the directions our national-preparedness programs were taking. World War II clearly proved the power of fixed-wing and flexible-wing aircraft as weapons of warfare. Also, during this war the Germans had brought to reality unmanned supersonic flight through their V-2 rockets. Thus, the big push in the immediate post-World War II era was for faster, highly maneuverable aircraft, heavier aircraft, the extension of supersonic flight to manned vehicles, and exploitation of technology for improved flexible-wing aircraft. From a national preparedness viewpoint, there was a sense of urgency in these demands. Although World War II was behind us, the Cold War had already begun, and mobilization problems associated with the war just ended were still vivid in the minds of those who had worked to solve them.

The heavy emphasis being placed on air vehicle speed and maneuverability posed new problems to airframe designers, particularly in the choice of materials that could withstand temperatures associated with the high frictional forces that would be encountered. High speed and maneuverability also equated to high stress on aircraft structures and the need for structural strength. All these factors added up to a requirement for new machine tools capable of sculpturing highly complex structural shapes from the lightweight materials suitable for operation in a high-temperature environment. As the Air Force's Webster explained:

> The Air Force development work [in NC] has been concentrated on continuous-path control rather than discrete positioning systems. This was because continuous-path control was deemed to possess potentiality in providing a highly flexible and accurate means for specifying and producing airfoil and other complex configurations peculiar to aircraft components. It was further believed that the solutions found to the problems associated with continuous-path control would apply equally well to discrete positioning.(6–1)

The Air Force agency charged with development of these machine tools was the Air Materiel Command (AMC), through its Manufacturing Technology Division at Wright-Paterson Air Force Base (WPAFB) in Ohio. Our recollection is that the Division embraced several branches, one of which was responsible for new machining methods. It was from this group that Parsons Corporation obtained its $200,000 contract, AF33 (038)–6878, dated June 15, 1949, to develop a machine for sculpturing helicopter rotor blades and aircraft wing panels through use of numbers stored on punched cards. Certainly this was a crucial turning point in the evolution of numerical control because public funds were brought to bear on a project that private industry was not ready to support in a serious way. From June 15, 1949 through May 28, 1951, Parsons received a contractual amount of $215,000 from the Air Force, and MIT, under its Memorandum of Agreement with Parsons extending from July 1, 1949 through January 31, 1951, incurred project-related expenses of approximately $80,000.

MIT's contractual relationship with the Air Force was highly satisfactory because the Air Force's civilian engineers who were assigned to monitor our work did just that: they looked carefully at what we were doing but never dictated how we should do it. We were also fortunate to have been associated with Air Force personnel who not only communicated well with the machine-tool and aircraft industries, but who understood the difference between the nonprofit environment of a university and the for-profit environment of the industrial sector. The organizational structure of AMC helped immensely in our contractual dealings with the Air Force. Career civil service employees formed a permanent core of project monitors, while the higher supervisory posts were usually held by officers at the lieutenant-colonel or full-colonel level. In contrast to the permanent civilian group, these men were assigned to a branch or division for limited tours of duty, a circumstance that had its advantages and disadvantages. Since the military had final say on budgets, and since our direct contacts were with our civilian project engineer, not his military supervisor, we had to rely on our project engineer to "carry the mail" for us at contract renewal time. As military personnel switched around, there was need to educate the newcomers about project goals—bring them up to speed. On the other hand, rotation had its good side; as time went on, we had an increasing number of military personnel in positions of higher authority who understood what we were trying to accomplish.

Servo Lab did, of course, have contact with the military supervisors of the project, although probably more so in the beginning than in the later stages. For example, Colonel Paul A. Brueckner of the Manufacturing Methods Branch took an active interest in the Air Force—Parsons contract, and toward the end of 1950 he led an effort to work out a plan with Messrs Parsons and Pease for its extension. As so often happens in research, the direction the project was taking in late 1950 was very different from the tasks that had been stipulated in the Parsons contract when it was initiated some eighteen months earlier. It is clear from a Pease memorandum of December 19, 1950, in which he summarized a series of meetings on contract renewal held at WPAFB, that Colonel Brueckner was trying to be helpful to the contractor while acting responsibly in behalf of the Air Force. Pease noted that Brueckner proposed a modification in the contract wording to bring it in line with the course of the present work, but that he "conditioned his proposal upon Parsons' representatives undertaking the job of 'reselling' the program to the Industrial Mobilization Division (particularly Colonel Brueckner himself)." This meeting also brought out the fact that Brueckner was sensitive to whether or not the machine tool builders and users were interested in the project. On this point, Pease wrote: "It was also proposed that Mr Sennett, the Air Force's project engineer, arrange to have machine-tool manufacturers and representatives from the aircraft industry visit Servo Lab to survey the project and to attempt to stimulate interest through a demonstration of the one-axis control."

Colonel Wilbur R. Carter also played an important role in moving numerical control into the marketplace. He was a strong advocate of getting industry to assume financial responsibility for NC development as quickly as possible. As indicated in an earlier chapter, at the time of the 1952 demonstration Colonel Carter thought it was time for the Air Force to withdraw from sponsorship of hardware development and to turn the technology over to industry for exploitation. In Carter's words: "My feeling, expressed in that meeting, was that we had, with government money, proven the feasibility and practicality of the idea, and I felt very definitely it was the responsibility of the machine-tool industry to pick it up and go from

William M. Webster

there. . ." (2–14). Carter was later assigned to the Pentagon as Chief of the Industrial Resources Division, Directorate of procurement and production, Office of the Deputy Chief of Staff, Materials. It was while he served in that capacity that the Air Force undertook its major machine tool procurement program, a circumstance that certainly did no harm to the plan to put the procured NC machines on aircraft company production floors rather than in storage. After "doing my four years of penance," as Carter described his Washington tour of duty, he returned to WPAFB where he once again played a background role in moving APT development ahead with industry support. However, it was still Carter's firm conviction that, while it was appropriate to use government money for proving the feasibility of new ideas, once proved, private investment should take over. He recalled:

> Boyd Bucy from Boeing, Seattle, was chairman of the Aircraft Industries Association's Committee for programming for NC and he came in to talk with me one day about an Air Force contract for getting a jointly developed and jointly acceptable computer language. . . .
>
> My feeling at that time was that it was primarily the responsibility of the aeronautical manufacturers . . . to get together and to undertake this sort of thing without . . . a government contract to pay the bill for it because it was their problem. Again, coming back to what I had said at the MIT conference [in 1952], we have proven the feasibility of the idea; the exploitations of it, I thought, were up to the machine tool manufacturers and the man who was going to use the tool, as well. (2–14)

So, having failed to get government support, the AIA put together a joint program between MIT and several aircraft companies to develop the Automatically Programmed Tool System (APT). The Air Force did contribute to this effort in part, however, through its APT contract with Servo Lab.

The three AMC civilian engineers at WPAFB who truly held the numerical control project together over the years were William M. Webster, Robert Jameson, and Max A. Guenther. Although Guenther was never assigned to the MIT contract as the Air Force's project engineer, he understood numerical control and saw its potential. He also enjoyed excellent relations with industry—particularly the aircraft industry. Webster was our WPAFB project engineer for 17 years, except for the short time he participated in the Air Force's advanced machine tool procurement program. During that interval, Bob Jameson took over the MIT assignment. In the 1950s, Webster had an important influence on the future of numerical control particularly with respect to the role NC tools should play in the bulk-buy procurement program. Although this procurement program was the responsibility of AMC's Industrial Facilities Division, Webster was a member of the working group that handled the details of the plan. The program was referred to by several names—the bulk-buy program, the long lead-time procurement, the "elephant" machine tool procurement, but finally and officially, the Machine Tool Modernization, Selective Augmentation and Replacement Program.

This program had its roots in the immediate post-World War II era. In 1946, the government had sold off large quantities of its machine tools as surplus property, thereby reducing its reserves and at the same time flooding the civilian market. The net effect was a drastic reduction in new industrial machine tool production. In 1950, at the start of the Korean conflict, the machine tool industry was operating at less than 10 percent of its peak World War II rate (6–2), a factor that seriously hampered machine tool procurements as the government geared up the country's production base for the conflict. As a result, the government's machine tool reserves (only partially replenished since 1948) had to be drawn down once again. However, to guard against a recurrence of this problem, the Air Force decided to initiate a new long lead time procurement program to replenish the stockpile and meet machining needs for future high-performance aircraft.

Originally conceived as a tri-service program at the Pentagon level, the initial objective was to acquire machine tools as a reserve against future mobilization requirements. NC played no part in the initial plan, but gradually thinking resulted in consensus that not only should NC tools be included in the procurement, they also should be put on airframe builders' production floors rather than into the reserve category. The long lead-time procurement program proved to be a big breakthrough for numerical control. Not only was a major market created for NC, but nonbelievers were compelled to address that market or be left behind. Furthermore, opportunities were created for a whole spectrum of users to gain hands-on experience in the use of the new technology.

Planning for the procurement spanned a five-year period from 1950 through 1954, with contracts being let toward the end of 1955. Final decisions on the kinds of tools to be purchased and how orders would be allocated rested with the Air Force, but representatives of the aircraft and machine tool industries were included in the decision-making process. A working group composed of WPAFB personnel worked closely with a task force appointed by the AIA. In addition, the machine tool builders were brought into these deliberations because, as Webster expressed it:"Before you could make. . . a procurement, you must have somebody willing to undertake the building of machines and. . . to stand back of the warranties to ensure that the machine and its controls will work (2–15)."

Besides specifying the kinds of tools the Air Force would buy and how they would be distributed among the aircraft companies, the WPAFB group also developed cutting tests which would be used as acceptance tests when tools were eventually delivered. During our 1969 interview, Colonel Carter expressed high praise for the group for their "aggressive and progressive" thinking about the bulk-buy procurement. To quote Carter:

> The group at Wright Field, I thought, was a heady group. I think that they had advice, counsel and assistance from the aeronautical manufacturers and from the machine-tool builders, but. . . they would have arrived at the same conclusion themselves. They reasoned. . . let's not look over our shoulder; let's look [ahead] at the day-after-tomorrow's problems and get ready for those problems. . . . (2–14)

Two real driving forces in the Air Force's decision-making process were Colonel Henry McDonald and Colonel Philip Kuhl, who worked closely together on a day-by-day basis. Providing internal support to McDonald and Kuhl were Lieutenant-Colonel Al Alcott; Messrs Max A. Guenther, William M. Webster, John Snyder, Harry Smith, Clarence O. Thomas, Jr; and Helen Haney. Supplementing this working group were several industry consultants to the Air Force, including Ralph Cross of Cross Machine Tool Company, Wallace Brainerd from Hughes Aircraft (later with Kearney and Trecker), and Walter Catey of the Ford Motor Company. Of this group, Colonel Kuhl and Mr. Thomas were extensively involved in the allocation process. The questions the Air Force group and its civilian advisors sought to answer were as follows: What kinds of machine tools should be bought? What kind, or kinds, of controls should they have? Should these tools be numerically or template-controlled? Or should they, perhaps, be template-controlled but with the potential for later NC conversion? Ultimately there was the question of whether the NC machines that were to be purchased should be stored or put to immediate use. The role played by industry in forming consensus on these matters is discussed later in this chapter. Suffice it to state here that of the $84 million allocated to the bulk-buy program for 500 machine tools, approximately $30 million went for 105 numerically controlled skin milling machines and NC profile milling machines plus their auxiliary controls. The rest of the appropriation covered the cost of 395 non-numerically controlled tools, including knee-type milling machines, vertical boring mills, broaching machines, turning lathes, and spar mills.

The impetus for placing the procured NC machines on production floors rather than stockpiling them appears to have come from the Air Force, although Webster, in our interview, made no claim to that effect. It was obvious, however, that he and Guenther strongly advocated this policy. Webster thought it would be foolish to store a new type of machine tool and later expect it to perform in a national emergency with operating personnel unfamiliar with this strange, new technology. He concluded that the fact that the machines were placed in production settings rather than storage was "mute testimony that the plan was accepted [by the aircraft manufacturers]." According to Webster, once the AIA task group saw the improved accuracies obtainable with NC over template control, in their check-outs of early machines, "they wanted numerically controlled equipment."

It is evident from this history that the government, through the forward-looking leadership of its Air Force, played a significant role not only in the creation of a wholly new technology, but also in ensuring the widespread acceptance and perpetuation of that technology in the marketplace. The government's role, however, did not end with its NC machine tool procurement program. It saw and responded to the need for a specialized high-level programming system for NC, lacking which the economics of NC would have been far less attractive. Furthermore, largely through the vision of our Air Force project engineer, William Webster, we were able to persuade our sponsor that machining by numerical methods was but one part of an entirely new manufacturing process, and that the government should underwrite research focused on computer-aided process methodology. The Air Force was indeed persuaded, with the result that the government came to play an important early role in getting the country's manufacturing base oriented toward what is now referred to as CAD/CAM—computer-aided design and computer-aided manufacturing. In the 1960s, the Air Materiel Command's Manufacturing Technology Division, through its Materials Laboratory at WPAFB, became the focus of the Air Force's activity in computer-aided design. For example, in mid-1966, when the US Department of Defense (DOD) needed a position paper on the Air Force's current and prospective involvement in the development of CAD concepts and systems, it quite appropriately turned to the Materials Laboratory, where the task fell to Webster (6–3).

The Role of the Servomechanisms Laboratory

Once our project staff had completely worked out the system design for a digital controller for continuous-path machining, they were confident that this technology had a future. Knowledge about digital computers and pulse-type servomechanisms was expanding so fast that it was inconceivable that the machine tool industry could remain untouched by these developments: recognition would eventually have to be given to numerical control. Sharing this view, our Air Force sponsor gave us strong encouragement to try to awaken serious interest in the new technology. Since the Air Force had no intention of forcing industry acceptance, the burden of educating machine tool users and suppliers fell to Servo Lab.

As an educational institution, MIT was in an excellent position to undertake the educational process. Dissemination of information was, and is, our business, and we were well aware of various techniques that could be employed to arouse interest. We used them all: demonstrations of the experimental machine for visitors; short, intensive courses for suppliers and users; on-site presentations to machine tool builders, airframe manufacturers and electronics organizations; cooperative programs with industry; publication of technical papers and reports; presentations at professional conferences; and most importantly, the inclusion in our research programs of a substantial number of graduate and undergraduate students as colleagues rather than pupils who would later enter industry equipped with experience in the new technology.

Almost from the beginning, there was considerable curiosity on the part of machine tool, aircraft and electronics manufacturers about what we were up to. From 1951 onward, a steady stream of visitors came to MIT to discuss the technology. The Lab's open-door policy of welcoming all comers produced, at the very least, an awareness that something unusual was in the making for machine tool control, even though initially there was plenty of resistance to the digital concept. Pease, in an internal memorandum dated June 6, 1951, described the prevailing viewpoint:

> [A representative of the Cincinnati Milling Machine Company] called me on Friday to say that people at Lockheed Aircraft were interested in a large four- or five-motion milling machine and had suggested that our control system might be applicable.
>
> Parenthetically, I believe the information channel started with Mr Edwards, AMC, our Project Monitor, suggesting the idea to us. We made no comment at the time although we did nothing to dampen his enthusiasm. Apparently, he then or through others at AMC passed the word along to Lockheed and from Lockheed it went to Cincinnati and now from Cincinnati it comes back to us.
>
> Before making any comments, I asked [the representative] what his views on the matter were and he was quick to state that he did not favor the proposal. His view seems to stem from a combination of risks and an opinion that he holds that our control system would not be suitable for production machinery. The risk factor is simply that Cincinnati has a proposal using their contouring techniques which they believe will solve the problem while they know nothing of our control system. We certainly can't blame them for taking the point of view that they have on that score. The other factor—that our control system is not suitable for production machinery—gave rise to some discussion. I maintain that that question cannot be answered at this time. There is no fundamental reason why we cannot use the control system on production machinery. He seems to have made up his mind that our control system would be for special prototype and model-making work.
>
> . . . He said that, although for this application he must talk our system down, he is still interested in our program and feels that ultimately it will find a place. (6–4)

Here indeed was a clear example of the way digital technology was being prejudged as unsuitable, even before it could be successfully demonstrated. However, Cincinnati Milling's prediction that NC would be applicable to prototype work proved to be an accurate forecast. For example, in 1961, when few if any could foresee a role for NC in the automotive industry, the Ford Motor Company became interested in the technology as a means of making prototype engine blocks (6–5). Only a few years later, the NC approach was standard practice at Ford's prototype-engine facility. The company had need for a flexible fabrication method that could quickly and inexpensively accommodate frequent design changes; NC provided that flexibility while also lowering potential for human operator error and reducing jig and fixture costs. We wish to add here that despite Cincinnati Milling's early skepticism about NC's applicability to production, they were always most supportive of the Laboratory. They provided drawings of several sample parts for our experimental machining programs and were well represented at our 1952 public demonstration. Nevertheless, it still took a lot of convincing to wean them away from template control to NC for production purposes.

Risk was frequently cited as a reason for industry reluctance to adopt digital control. A related factor was tradition: in fulfilling orders for machines, the builders preferred to cling to time-tested, proven methods. It is possible, of course, to be empathetic to the conservative position we initially encountered in the industry. Machine tool technology had evolved over many decades to a highly refined state in which machining to a precision of fractions of a thousandth of an inch was standard practice. Also, substantial investment had been made in template control as a means of improving productivity, which it had in fact done. Prior to completion of the experimental machine, and for some time afterward, we had a major educational job to do on the issues of accuracy and reliability. The closer machine tool designers and users were to high-accuracy requirements (machining to one-thousandth of an inch or better), the more skeptical they were that digital control, as a concept, would ever achieve the accuracy they demanded. In light of their experience with instrumentation and control, this position was justified; what was missing was an appreciation that with the emergence of digital technology, the accuracy issue became a non-issue.

On the subject of reliability, there was similar suspicion—indeed disdain—for any control system that used vacuum tubes. One can hypothesize that part of this negative bias came about through individual experience with consumer electronics—radios and television sets that frequently needed service and repair in those days, mostly because of vacuum tube failure. The remarkable progress made by manufacturers in the 1950s in producing highly reliable vacuum tubes for military and computer applications had not yet penetrated the consciousness of mechanical designers.

Thus it was that, until we had a digital control system we could demonstrate and experiment with, we were able only to make predictions of reliability and accuracy during discussions at the Laboratory and at outside presentations. However, once we had an operating system, about March of 1952, we could be more assertive in our claims. It was at this time we began to solicit airframe builders for drawings of parts included in their vehicles so that we could gain firsthand experience in everyday industrial parts fabrication and begin to accumulate tangible evidence of NC's technical feasibility. Gradually, through these efforts, thinking began to incline our way. For example, after a mid-1952 ten-day tour of industry facilities around the country, McDonough and Susskind reported active interest in our system and a willingness to supply us with experimental jobs. Here are some excerpts from their 1952 trip report (6–6):

June 23 Aeroproducts Division of General Motors Corporation, Dayton, Ohio
The acute need for templates and masters had prompted Aeroproducts to contact us. It was agreed that the MIT machine is directly applicable to making templates and appropriate drawings for sample jobs would be sent to MIT for manufacture. .. .

June 24 Boeing Airplane Company, Wichita, Kansas
Arranged for receiving 27 forgings of the Forward Drag Strut. . .

June 25-26 Consolidated Vultee Aircraft Company, San Diego, California
Dr Short (Special Projects and Studies) mentioned he is preparing a talk on digital techniques before the Board of Directors. It was agreed that MIT would furnish to Dr Short illustrative material, such as a tape, a finished part and literature.

June 27 Douglas Aircraft Corporation, Santa Monica, California
General discussion of numerically controlled milling machine, demonstration of sample pieces, discussion of program at MIT. . .

June 30 General Motors Corporation, Detroit, Michigan
Agreed that MIT would make, as a sample job, a master of the casting core for one of the GMC torque converters. This is a difficult job, both for GMC and MIT and will test our machine in the area of three-dimensional complex shapes.

July 1 NACA, Lewis Flight Propulsion Laboratory, Cleveland, Ohio
The MIT numerically controlled milling machine was described and discussed.

July 2 McDonnell Aircraft Company, St. Louis, Missouri
Discussed the application of numerical control to airplane model manufacture. It is clear that numerical control offers advantages, provided the complex contours do not involve excessively complex programming procedures. . . .

McDonnell expressed an interest in applying numerical control to other production problems. It was agreed that McDonnell would send us drawings of typical parts for experimental planes. . . .

As is apparent from the above excerpts, the project was beginning to gain the attention and cooperation of the aircraft industry, but the burden was still on us to prove the worth of the new technology.

The September 1952 public demonstration of NC was an especially valuable experience for the Servo Lab staff because it revealed areas where more educating had to be done. While the meeting provided excellent project visibility, it was obvious that industry was not champing at the bit to pick up the concept and run hard with it. Much work was still needed on the economic aspects of NC. Moreover, we needed to reach industry decision-makers to convince them to invest in their own in-house R&D programs, and we had to provide electronics and mechanical designers with the details of numerical control from a *system* perspective. This latter job was a critical one; if NC were to become a reality, the electronics and machine-tool specialists would each have to know the capabilities and limitations of the other's discipline. Consequently, after 1952 our site visit strategy with machine tool manufacturers, electronics houses and aircraft builders was intensified; visits continued under forced draft for the next year and a half. This information dissemination program was coordinated by the Lab's Richard Lawrie, who spent full time on the effort. It generally took the form of an oral presentation on NC augmented by a twenty-minute, narrated motion picture film of the experimental machine tool system in action. This approach ultimately paid off handsomely for at least two reasons: one manufacturer, the Giddings and Lewis Machine Tool Company (G&L), decided to invest in a digital director system, designed to be compatible with an electronic playback control system then under development by the General Electric Company for G&L's five-axis skin milling machine. The broader return, however, came as users and suppliers became more familiar with the concept; as a result, numerical control entered more frequently into discussions between the aircraft manufacturers and the Air Force concerning its long lead-time machine tool procurement program.

To meet the growing need for technical information about NC, Servo Lab staff offered two intensive two-week courses for electronic and machine designers during the summers of 1954 and 1955. These were internal MIT educational programs and were in no way connected with our Air Force sponsor; they were administered by MIT's Summer Session Office in accordance with its prevailing format. The 1954 course brochure stated:

> This special Summer Program is directed toward persons who desire to obtain a broad view and understanding of the technical aspects of numerical control as they apply to machine tools. It will be especially significant to engineers contemplating the application and use of such equipment and therefore desiring direct experience in the programming and use of a numerically controlled machine tool. Since the number of registrants must necessarily be limited, preference will be given to applications from persons now engaged in the design or application of automatic machine-tool equipment or anticipating entrance into this field.

Classes were held from 9:00 AM to 5:00 PM, Monday through Friday, with classroom presentations in the morning and programming and metal-cutting sessions on the experimental milling machine during the afternoon sessions. Sixty attended the course in 1954 and another 60 in 1955. Each participant received a 236-page package of notes on 16 technical topics related to NC, and a Certificate of Completion was awarded on completion of the course. The rosters of course attendees, found in Appendix C, reveal a broad cross-section of industry, university and government representation. Figures 6.1 and 6.2 are photographs of the two classes.

FIG. 6–1. Participants in the 1954 special summer program on numerically controlled machine tools.

FIG. 6–2. Participants in the 1955 special summer program on numerically controlled machine tools.

First Row (l to r): G. A. Harris; W. H. McKnight: J. E. Stell; E. F. Gallagher; G. S. Glinski; G. E.Lingenfelter; J. F. Reintjes; E. L. Juell; D. H. Ross; H. W. Brainard; F. J. Yannotti; R. L. Curtner; A. L. Consalvi; D. C. Dick

Second Row (l to r): J. H. Brown; J. J. Sennello; H. B. Buchanan; W. W. Wilder; J. A. Wilkening; V. E. Tice; E. F. Carlberg; R. G. Chamberlain; H. P. Grossimon; W. H. Lynch; J. Fondrk; J. F. Salatin; W. A. Fletcher; J. K. Webb; J. J. Finden; R. W. Lawrie; P. W. Winkelman

Third Row (l to r): F. X. McGarrigle; A. H. Stillman; P. Tarver; M. I. Schneebaum; P. J. Issacs; D. C. Caulkins; W. W. Woods; L. S. Peck; J. Feiner; L. V. Griffin; A. L. Keller; E. R. Beck; T. A. Wetzel; R. D. Rutt; J. O. McDonough; P. C. DeLuca

Fourth Row (l to r): L. E. Walker; R. S. Litterst; C. H. Laommel; C. F. Carter; R. R. Hawkins; J. Meess; C. F. Erikson; R. A. Wade; R. W. Pfaff; N. Zlatin; S. A. Duran; R. Bartsch; W. M. Webster; A. K. Susskind

FIG. 6–3. Participants in the 1957 special program on programming for numerically controlled machine tools.

First Row (l to r): V. Benatar; T. M. Rothrock; A. L. Eskelin; R. Kuehn; E. C. Goudreau; D. H. Ross; J. B. Rankin; R. E. Stalcup; W. J. Kinney; M. J. Lacy; J. F. Reintjes (MIT)

Second Row (l to r): E. B. Dohner; A. D. Wheless; C. E. Butzberger; C. W. Trowbridge; A. H. Blacker; L. H. Bradshaw; R. A. Bennett; B. E. Forsman; G. R. Boucher; R. I. Hagar; D. T. Ross (MIT); N. Melissas (MIT)

Third row (l to r): F. W. Olsen; H. F. Robinson; K. C. Sanderman; H. R. Montgomery; L. E. Ording; O. R. Evans; J. Albert; S. S. Messick; C. E. Robinson; R. C. Van Buskirk; R. H. Gregory (MIT)

Fourth Row (l to r): L. R. Mhyre; R. W. Miller; B. A. Richardson; L. E. Walker; M. Lehrer; E. E. Graham; S. T. Byars; W. P. Robertson; H. A. Leone; A. D. Baumbach; H. M. Lakin; A. Siegel (MIT)

Fifth Row (l to r): C. F. Quick; K. C. Roye; S. L. Samuels; R. O. Young; C. O. Parratt; R. S. Watson; A. D. Sharp; W. M. Webster; D. H. Brown; H. P. Philpot; C. B. Young

Six Row (l to r): F. Belanger; J. J. Childs; F. W. Mowry; V. M. Werbach; E. A. Green; J. A. Horn; J. B. Tyson

The following participated in the instruction of the Course but were absent when the picture was taken: T. Atwater; D. F. Clements (MIT); D. C. Dick; H. P. Grossimon; H. Pople (MIT); A. K. Susskind (MIT); K. G. Wood.

So convinced were we of the value of this two-week intensive course format that we used it again in 1957 to acquaint industry representatives with our work on automatic programming. Once again, industry responded enthusiastically: 57 people from 14 aircraft companies and nine suppliers participated during the period March 25 through April 3, 1957. This class is shown in Fig. 6.3; their names and affiliations also appear in Appendix C. However, the tone of this course was quite different from that of its two predecessors. Whereas in 1954 and 1955 information flow was mostly from instructors to participants, in 1957 the participants brought with them substantial knowledge about NC—by that time, plans were well underway for receiving the numerically controlled tools purchased under the Air Force's bulk-buy procurement. There was extensive class discussion both with the instructors and among themselves. It became apparent that the industrial sector was allocating a substantial amount of manpower to NC, and organizations with an interest in the technology were acquiring considerable in-house competency. For example, we were able to augment our instructional staff with the addition of Kenneth G. Wood, then Assistant to the Manager, Facilities Engineering at Boeing, Seattle. Ken was truly a tower of strength on NC matters, not only for his company but for the industry as a whole. He served with distinction as a member of the Aircraft Industries Association's Subcommittee on Numerical Control and played an important role in ensuring the inclusion of NC machines in the government's machine tool bulk-buy program.

It became evident during the 1957 course that there would soon be a shortage of programmers skilled in NC programming techniques. A special need existed for programmers who not only understood numerical control but also had some knowledge of metal cutting, cutting tools, and jigs and fixtures. Therefore, as part of the course, the participants and instructors prepared an outline to serve as a guide for industrial organizations planning to set up company training programs in numerical control, with emphasis placed on familiarization with the concept, programming, and data processing. This outline is reproduced in Appendix D. Although there is no record of how or to what extent this document was actually used in industry, it did serve to focus course participants' attention on the need for programmers with new kinds of skills who would function effectively in the emerging NC environment.

While Servo Lab's central purpose in offering the 1957 course was to acquaint organizations interested in NC with our recent work on automatic programming, we also hoped to stimulate outside support for a cooperative plan that would lead to a unified, industry-wide automatic programming effort, rather than to individual, fragmented efforts. As part of the course, we were able to demonstrate and discuss a rudimentary version of the Automatically Programmed Tool (APT) system of programming for NC currently operating on MIT's Whirlwind I computer. The staff's viewpoint on automatic programming for NC was received enthusiastically by the participants and undoubtedly had an impact on their thinking and on plans for coping with long-range programming matters, as evidenced shortly thereafter by the AIA's agreement to undertake a joint, aircraft-industry-wide APT programming effort, with Servo Lab as overall coordinator. Our coordinator role was strongly supported by our Air Force sponsor, and the AIA–MIT collaboration culminated in the achievement of an operational programming system, designated 2D-APT-II, which is described in detail in Chapter 4.

As the day approached when an early operational version of the APT system was ready for release to industry, the AIA, the Air Force and Servo Lab agreed that a formal pronouncement was the appropriate way to publicly recognize such a substantial accomplishment; in addition, it would provide an opportunity to attract the interest of potential users. Plans were made for a USAF–AIA–MIT-sponsored press conference to be held at the Institute on February 25, 1959. In response to our invitation (see Appendix E), 49 representatives of the technical press and the news media—writers, reporters and photographers—came to the one-day meeting. Each received a kit containing a dozen items of information, including a description of the APT concept of programming; explanations (in technical and nontechnical language) and diagrams of how APT worked; important milestone dates in NC and APT development; a news release from the Air Force; biographical sketches and photographs of the Air Force, AIA and Institute speakers who were to address the conference; a set of relevant NC and APT publications; and photographs of numerically controlled machine tools. In addition, each member of the press received a memento of the occasion in the form of a souvenir ashtray, sculptured on MIT's experimental NC milling machine operating under an APT-generated program. The souvenirs had been machined ahead of time, but as part of our demonstration, one was sculpted during the conference.

The press conference at MIT succeeded in gaining a broad exposure for the APT system of programming for numerical control. Write-ups in the technical press detailed APT's features succinctly in language that ranged from highly technical to a level suitable for the non-expert but technically oriented person. The popular press, however, tended to embellish its stories with eye-catching descriptions to arouse reader interest, often seizing on the term "robot,"or introducing ominous implications regarding the human operator. For example, the *Boston Globe* headlined its APT system story: "HUMAN ELEMENT GONE. ROBOTS RUN MACHINES. . . " On the other hand, *The New Yorker* magazine saw humor in a particular news release, as witnessed by the following reprint and comment which appeared in its pages:

> *Cambridge, Mass. February 25*—The Air Force announced today that it has a machine that can receive instructions in English, figure out how to make whatever is wanted, and teach other - machines how to make it.
>
> An Air Force general said it will enable the United States to "build a war machine that nobody would want to tackle."
> Today it made an ashtray.
>
> ©*San Francisco Chronicle*
> Reprinted by permission

Quipped *The New Yorker:* "In a sulk, probably."

Societal Issues

Servo Lab's presence in the field of numerical control inevitably drew staff members into discussions—rampant in the early 1950s—on the controversial subject of automation, a newly coined term for automatically controlled processes. In fact, so new was this terminology that there were differing opinions not only about what it meant but how to spell it. (Some held out for automation; others maintained it should be automatization.) The advent of the digital computer and servomechanism technology, and their merger in the development of real-time digital controllers for machine operation, were undoubtedly responsible for this upsurge of public interest—particularly with respect to their societal implications. MIT, as a leader in research and development of the principal technological ingredients—computers, feedback control, and automatic materials-handling equipment—was frequently called upon to provide expert commentary on these subjects. Servo Lab personnel spoke to the technologies behind numerical control, but more often than not discussions would veer off in the direction of just how the general field of automation would influence jobs and productivity. Labor organizations in particular were apprehensive about job displacement and retraining of workers in light of increasing use of automatic manufacturing equipment, which was seen as diminishing the need for highly skilled machine operators.

The positions of Gordon Brown and the author on these matters were set forth in several talks given during the 1950s (6–7 through 6–12). Our position was, and still is, that technological innovations have contributed positively to the common good in the past and have the potential to do so in the future. We acknowledge that a consequence of technological innovation is that jobs will change, and some may even be eliminated (a most disruptive matter for those whose jobs disappear), but that in the aggregate more jobs will be created than will be lost. Our further view was and is that both labor and management have responsibilities in light of changing job patterns: management has a societal responsibility to provide for displaced workers through retraining and other procedures, and workers have a responsibility to recognize the need for a flexible, adaptable approach to technological innovation in order to ensure the economic welfare of their company. In our presentations we continually stressed these points, and also the need for *visionary planning* by management and labor, *working together*, to accommodate technological change. As the author stated in an address to representatives of the United Steel Workers of America: "I should like to rule out 'Men versus Machines'. . . May I propose instead that we take as our platform: 'Machines to Serve Men Better'." (6–9).

In looking back to the 1950s, and the many talks given by Brown and the author on the societal implications of automation, we would like to believe we had at least a small positive impact on the thinking of management and labor about nontechnological issues related to technological change. Frequently we stressed the need for life-long learning. Perhaps it is possible to see a common thread linking those talks, the outline for a retraining plan developed with the help of Boeing employees and others during our 1957 NC course, and the 1983 Agreement between Boeing and its union, the International Association of Machinists and Aerospace Workers, AFL/CIO. In their collective bargaining agreement (6–13), both parties stated that "it is to their mutual benefit and a sound economic and social goal to utilize the most efficient machines, processes, methods, and/or materials" in order that the company might compete effectively in the marketplace and "thereby provide economically secure jobs for its employees." The agreement also stated that it is the "Company's policy to assure that training is available for its employees so that they may have the opportunity to acquire the knowledge and skills required by the introduction of new technology." The document went on to describe the new technologies as being industrial robots, flexible manufacturing systems, CAD/CAM, and graphite composite automation, and to prescribe procedures for implementation of training programs. A feature of the contract was that it covered employees in a layoff mode as well as active employees. The agreement came at a time of heavy layoffs; some 7000 employees in the Greater Seattle area were out of work because of slowness in Boeing's businesses—mostly the commercial airliner business. Subsequently, all but two hundred workers were called back, many of them now in different jobs, which they were able to fill because of retraining programs (6–14).

Graduate and Undergraduate Students

The numerical control project takes on an added dimension when one considers the role it played in educating and shaping the careers of the one hundred fifty or so students who participated in the project over its twenty-year life span. For these students it meant firsthand contact with an innovative engineering research project and making genuine contributions to its progress through their undergraduate or graduate thesis research. Some later joined our staff and became leaders of important segments of the project, and at a remarkably young age, too: most were in their late twenties and early thirties. Others, having gained their internship training at Servo Lab, carved out illustrious careers for themselves in their specialized fields, both in industry and education. Still others chose the entrepreneurial route; they founded new business enterprises based on the technologies they had helped create. We submit that the combination of classroom instruction and an engineering research experience, gained in an environment that engenders new ideas and creates opportunities to develop and exploit them, is the kind of educational experience that contributes strongly to the forward progress of our nation and the world. Numerical control is but one product of such an environment.

In the paragraphs that follow are glimpses of the careers pursued by a few of the people who, as students, worked on our project.

Al Susskind describes his recruitment by Servo Lab as follows:

> [In the fall of 1949] I was taking a course in servos from Bill Pease; he came up to me after class and said: 'Say, may I talk to you about coming to work in Servo Lab?' I think it was early January 1950—it was at the very end of the semester. I talked with him for a while and I talked to Al Sise [executive officer of Servo Lab]. So I came to work in February. (6–15)

At the time he joined the Lab, Susskind was a research assistant and was finishing his Master's degree thesis on a digital computer topic of great interest to Jay Forrester. Forrester had just completed his review of the NC project for Gordon Brown and had recommended to Pease that a full-time digital circuit designer be added to the project roster. He put forward Susskind's name as the right man for the job. Pease did the recruiting, and Susskind became the principal designer of the electronics for both the experimental NC milling machine and the Giddings and Lewis digital director. Later he became Assistant Professor of Electrical Engineering at MIT, teaching subjects related to digital and analog electronics; still later he served as head of the Electrical Engineering Department at Lehigh University.

Jay Forrester became the leader of the Laboratory's Whirlwind I computer project after serving as a graduate research assistant and performing his MS thesis research under the supervision of Professor Brown. Although never formally associated with the NC project, he made important contributions through his early critique of its direction and his suggestion for the design of a critical circuit that led to a major simplification of the digital director system. Forrester has remained at the Institute throughout his distinguished career; he is presently Germeshausen Professor of Management at the Sloan School of Management, where his work in the application of computation and control to management problems has gained him recognition as a world leader in the field of industrial dynamics.

Bill Pease's career is equally distinguished. After receiving the SB and MS degrees in MIT's Cooperative Course in Electrical Engineering, he became a staff member of Servo Lab. He was later appointed Assistant Professor of Electrical Engineering and became director of the Lab when Brown relinquished the post to become head of the Electrical Engineering Department. In 1953, Pease resigned his faculty position to pursue a career as an engineer and manager in industry. He has distinguished himself over the years in these capacities at Ultrasonic Corporation, Feedback Controls, RCA, and the Raytheon Company.

Jim McDonough and Doug Ross chose different routes from those cited above. After serving as a graduate research assistant and completing his MS thesis in Servo Lab, McDonough joined the Servo Lab research staff and later succeeded Pease as leader of the NC project. Upon completion of the Giddings and Lewis digital director system, McDonough was encouraged by Edgar McFerren, vice-president of G&L, to form a company (with some financial backing from G&L) to manufacture the digital directors G&L had sold to the government. McDonough, together with seven members of his Servo Lab group—Donald Dick, Ernst Herzberg, Herbert Grossimon, Henry Kilroy, Richard Lawrie, Gerald Moore, and John Morin—founded the organization known as Concord Control, Inc. and thus transplanted their Servo Lab experience into industry.

Doug Ross was a research assistant in the Mathematics Department when he first came to Servo Lab. As a young man in his twenties, he successfully created the APT system of programming for numerically controlled machine tools and brought about the joint programming effort with the aircraft industry that led ultimately to acceptance of APT as a world standard. Having carried through that task, Ross saw the need for a high-order computer language as the essential ingredient of computer-aided design of NC systems. He turned his attention to the development of a language that would be both easy to use and very powerful in its ability to assimilate user instructions and to execute them without human intervention. His recognition of this need led to the development of the AED language discussed in Chapter 5. On completion of the computer-aided design project, Ross and three of his group members—Clarence G. Feldmann, Jorge Rodriguez and Jerry Welch, all former MIT research assistants—left the Institute to found SofTech, Inc., a software development house that continues to make important contributions to the military's software system needs.

The careers of two other project staff members, Steven A.Coons and John E. Ward, will serve to further illustrate the influence our Lab environment had on the development of numerical control. As discussed in the preceding chapter, Professor Coons of MIT's Mechanical Engineering Department became part of the Servo Lab project when we embarked upon the computer-aided design phase. Unlike the others, Coons did not come up through the MIT student ranks. He was a top-notch designer who was able to foresee the potential of computers in the mechanical design process. As noted earlier, in his book *Numerical Control, Mathematics and Applications* (5–6), Professor Bezier devotes a section to Coons' mathematical descriptions of curved surfaces, known as Coons patches. Through his consulting work with the Ford Motor Company, Coons demonstrated the applicability of his mathematical descriptors to the design and styling of automobile bodies, using digital computers driven by interactive graphical display terminals. Undoubtedly he was among the first to excite the interest of automotive manufacturers in computer-aided design for automobile body styling and other applications.

John Ward did come up through the MIT student ranks, beginning his career as an MIT undergraduate in 1940. At the end of World War II, he joined the Servo Lab staff as a research assistant, while completing the requirement for his MS degree. During his long career at the Lab, Ward led several research projects. It was he who first perceived the enormous potential for interactive graphical displays—if they could be produced at a reasonable cost. He therefore set a course that would lead eventually to the development of low-cost displays, first by getting his contacts in industry to recognize that need and then by working closely with them to find solutions. Ward's ARDS display terminal, described in Chapter 5, was an early breakthrough in low-cost graphical displays, and led in time to even simpler, lower cost display systems. Although today's graphics displays differ in form from Ward's original, they are within the financial reach of personal computer owners, undoubtedly because of his pioneering efforts.

These colleagues who contributed so importantly to the development of numerical control and computer-assisted design are cited here not only because of the importance of their contributions, *per se*, but also because of what can be called the educational amplification factor. It was they, together with a few faculty members, who created an academic research environment that influenced more than one hundred fifty students and staff members, all of whom eventually went into industry or teaching professions where others would benefit from their project-acquired skills. In addition, the Lab staff's influence extended to at least another eight hundred persons who became affiliated with us in one capacity or another—as visiting faculty, as collaborators in the cooperative APT development effort, as CAD project visiting staff, or as participants in the three intensive courses on NC and APT and in the three conferences on AED. Then also there was the steady stream of visitors, numbering many hundreds, who came to Servo Lab for informal discussions or to witness a demonstration of the experimental machine.

The Role of Industry

In the evolution of numerical control, the role played by industry was crucial. The prospective users of the technology—mainly the aircraft manufacturers—initially had the job of understanding the concept of NC and its potential for positive contribution in their manufacturing environments; next came the job of developing specifications for the kinds of machine tools they wanted; and finally, they had to find or retrain personnel to operate the tools after orders had been placed. The supplier segment—principally the machine tool builders and the electronics companies—had the task of supplying NC machines and related digital controls to meet user specifications. This posed a major challenge. At that time, machine tool builders had few in-house staff trained in digital computation and control. They were well staffed with mechanical designers and mechanical engineers sufficiently skilled in the electrical domain to handle electromechanical power-drive problems, but these people were not conversant with the more advanced digital and servomechanism technologies required to design continuous-path numerically controlled machines.

Before the advent of numerical control, machine tool builders had already raised automatic machine tool design to a state of high refinement, principally through the expertise of their mechanical engineers and designers. Substantial technological progress had been made during the 1940s and early 1950s as a direct result of the huge demands placed on the industry during World War II. For example, in the realm of special purpose metal-forming machines, stand-alone multiple-stage-die machines were in use for automatically stamping and forming metals for a wide variety of applications. Precision special purpose metal cutting machines also existed, and in high volume situations this type of machine was often concatenated and linked by automatic materials handling equipment to form production lines—the so-called "Detroit, automation" concept used in automobile manufacturing. Metal-cutting machines employing template control were used to replicate parts in small-lot production runs of a few hundred pieces, particularly in the aircraft industry.

The machine tool builders were certainly not oblivious to electronic machine tool control in the late 1940s and early 1950s. The major companies operated R&D laboratories, but motivation for moving electronic control technology from the laboratory to production floors was lacking, undoubtedly through their inability, at that stage of development, to meet high-precision requirements, and the absence of profitable markets for machines with the accuracies attainable. However, the picture changed dramatically with the advent of numerical control; digital technology was able to meet high-precision requirements and its acceptance by the aircraft industry created a market.

The aircraft companies approached numerical control from a different viewpoint. Airframe builders were eager to know what NC could do to reduce machining costs and shorten product delivery times. Both factors were becoming increasingly important in the 1950s as machine tools became more complicated because of their size, and as jig and fixture costs and setup times mounted because of the increased complexity of parts in the newer aircraft. Fortunately, when the time came for the aircraft industry to make a critical assessment of NC, it had qualified personnel within its ranks to make the assessment. Digital computers for solving mathematical problems associated with airframe design were finding their way into the major companies, and at hand were employees who knew enough automatic control theory and digital computer technology to do a competent evaluation job. Through its industry-wide trade association, the Aircraft Industries Association, the industry had a salutary influence on the future of NC. When the Air Force made its decision to proceed with the long lead-time procurement program, the AIA was in a position to respond quickly to the Air Force's request for an aircraft industry task force to advise on the selection of appropriate tools for inclusion in the procurement. The task force reported to the AIA's Airframe Manufacturing Equipment Committee, one of several committees reporting to the Manufacturing Committee. Among the AIA task force members, Webster, in our interview, recalled Ken Wood of Boeing as being "one of the best-oriented people," as well as M. C. Copold of Convair, San Diego, and Ken Coleman of Lockheed, but he hastened to add that there were others, too, from other companies.

Before making recommendations on the kinds of tools and controls that should be procured, the AIA task force mounted a thorough study, not only of industry needs but also of the comparative capabilities of relevant technologies. Care had been taken to include the machine tool builders in these deliberations. Needless to say, there was intense competition among the builders, not only for inclusion of their machine tools in the procurement, but also for the kinds of controls to be supplied. As task force meetings continued, discussions turned inevitably to the wisdom of including numerically controlled machines in the procurement. The rationale of the Air Force's program was such that NC could be easily justified. The goal was to ensure an adequate production capability in the event of a future national emergency and to procure a supply of the advanced tools that would be required to solve future machining problems. As might be expected, there was considerable foot-dragging by builders who had no NC machine tools ready to market, and aggressive promotion by those who did. Builders with some knowledge of and capability in numerical control technology (notably Giddings and Lewis and Kearney and Trecker) favored its inclusion in the purchase; others, who had nothing to offer, disagreed. For example, Lockheed and Cincinnati Milling, who were teamed in an existing Air Force contract to develop a large five-axis template-controlled skin milling machine, favored template control. However, as discussions progressed, it became clear that numerical control was gaining momentum. A consensus gradually formed that at least some of the machine tools in the procurement should have an NC capability. The recommended number was 63 NC machines out of a total of 105. However, when the Air Force drew up procurement contracts, a proviso was included that template machines must be capable of later conversion to numerical control. This action undoubtedly signalled that while the Air Force looked with favor on NC, it was not about to force the issue with the industry.

None-the-less, the proviso did have a telling effect. According to Webster: "One day a meeting was called [at WPAFB] at the request of the aircraft industry's task force. . . . These people came in and recommended without qualification that for the aerospace industry, all profilers, all skin mills, and all spar mills, all of them, be numerically controlled, that none of them be template controlled as we had previously established a contract for." (2–15) This was a major recommendation. As a result, the aircraft industry formally recognized NC as a viable technology and created just the market needed by the machine tool builders to get them seriously involved with NC. Thus, the original procurement of 63 NC machines was augmented by the conversion of 42 profile mills from template control to numerical control. Four different numerical control systems were included in the procurement, as well as six digital directors which were needed to support two of the NC system types.

The bulk-buy program gave further impetus to numerical control because of the way orders for NC machines were distributed and the manner in which the machines were parceled out. Most of the major aircraft companies were supplied with machines, but the aircraft builders' *suppliers* also received NC machines. Webster made a point of this (2–15), citing the impact these machines had on the die-sinking business. Through another branch of AMC's Manufacturing Division, several NC profilers went to organizations such as Kaiser Industries, Alcoa and others for use in their die-sinking businesses. The fact that numerically controlled machines were allocated to many different organizations ensured widespread usage of the technology, and the fact that several machine tool builders received orders for NC machines undoubtedly promoted a spirit of healthy competition among those organizations.

With orders in hand for numerically controlled continuous-path machine tools, the builders began the task of finding electronic engineers skilled in the areas of digital electronics and servomechanisms to design and fabricate the controls. This would prove to be a formidable job. In the 1950s, the computer and automatic control fields were expanding in leaps and bounds, and the prestigious jobs for trained engineers in these disciplines were with the electronics and computer companies, not in the machine tool business. Consequently it was difficult for the machine tool builders, with their conservative image and their mechanically oriented personnel, to persuade the bright, aggressive, young post-war-educated electronic specialists that they had a future in the machine tool business. Nevertheless, hiring an in-house group of digital control engineers was deemed the best solution. Giddings and Lewis, an early leader in NC, tried this approach without success. The obvious group to hire was MIT's Servo Lab NC group that had developed G&L's digital director. However, the group was unenthusiastic. We recall that they had doubts about the interest of bright young electronics specialists in working in the relatively isolated environment of Fond du Lac, Wisconsin, under the leadership of mechanical engineers and machine designers.

Another possibility for the builders to establish competency in digital control was to enter into a cooperative agreement with an electronics company. The advantage of this approach was that it provided essentially an instant presence in the electronics field; the disadvantage was that the machine tool organization would lack complete control over the electronics segment of its business. McFerren of G&L explained that he had tried to work out an arrangement with General Electric but was not successful (4–4). General Electric was the natural organization for G&L to turn to since GE was already a supplier of some G&L controls. However, GE decided against a cooperative agreement with G&L. According to McFerren: "The people at General Electric pretty much downgraded the [G&L's] original work," and he ventured the thought that "maybe they were hesitant to pick up someone else's work." He was referring to the digital director Servo Lab had designed for G&L.

McFerren also held a series of conferences with IBM, but it too declined to enter into an agreement. IBM gave the proposition serious study but concluded that furnishing machine tool controls was not a good line of business for the company. IBM did not downgrade G&L's work and the potential of NC; it simply felt it should stay with the computer end of the business and confine its activity to machines that processed NC information, but stay away from servo controls.

Giddings and Lewis finally got into the electronic controls end of the NC business by pursuing a new tack with the Servo Lab NC group. The proposition which the group finally accepted called for the establishment of a separate corporate entity, as a subsidiary of G&L, in the Cambridge,

Massachusetts, area. The subsidiary company would fabricate the G&L digital directors ordered by the Air Force under the bulk-buy program, but it would also be free to create and exploit its own markets for digital processing and control systems. McDonough was the group's spokesman during the negotiations with G&L; when the agreement was concluded, he became president of the new company, Concord Control, Inc., located in Brighton, Massachusetts. As noted earlier, seven Servo Lab NC staff members left to join McDonough's company. In this turn of events we see a familiar pattern, repeated over and over again in the free world: a group of bright, enthusiastic young people developing a highly innovative process or product and wanting to capitalize on their innovation by establishing their own enterprise. It is this entrepreneurial spirit that has contributed so greatly to the steady expansion of our nation's business base and to the steady growth of our economic system.

Other machine tool companies followed similar procedures in their attempts to gain a foothold in digital electronics. For example, Cincinnati Milling chose to establish a relationship with a British organization, Electrical and Musical instruments (EMI) a relationship incidently that did not endure, for it soon began to manufacture control systems of its own; and Kearney and Trecker joined hands with Bendix (Detroit) as its supplier of electronic equipment. Bendix's 1956 brochure describing its tape control system emphasized that it was "an automatic system that reduces the time and cost of producing complex three-dimensional machined parts. . . . The system controls the manufacturing operation from blueprint to finished part by means of a punched paper tape. No templates, cams, models or fixtures are required, and, as a result lead time and cost are reduced." In its price confirmation quotation to Kearney and Trecker (6-16), Bendix specified its tape control system in two parts: item 1 included a machine control unit, three servo drives, three feedback instruments, and a hydraulic power supply, at a per-package price of $49,500 in lots of 22; item 2 covered the tape preparation equipment, at a cost of $54,910 in lots of 5 or more. The Bendix tape control system was part of the NC profile milling machine Kearney and Trecker furnished to the Northrop Company under the Air Materiel Command's procurement program. Figures 6.4 and 6.5 illustrate two NC milling machines in Kearney and Trecker's product line.

Bendix's Research Division began work on numerical control systems in the early 1950s under the leadership of Dr Albert C. Hall, a former MIT Faculty member and staff member of the Servomechanisms Laboratory. According to Hall, Bendix's jet engine division had a requirement for a three-dimensional cam for controlling fuel injection to its jet engines. The task of developing a machine-tool control system for cutting the cams was undertaken by Hall's group, which chose to exploit digital techniques for that application. This experience led to a contractual relationship between Bendix and the Glenn L. Martin Company at Baltimore, in which Bendix developed a numerical control system for a small-size three-axis milling machine, one of the first, if not the first, commercially manufactured machine tools of its kind.

Cooperative arrangements between machine tool builders and electronics organizations proved to be only stop-gap measures. The major machine tool companies embarked on long-term recruiting programs, and now most have their own in-house staff of computer and feedback control engineers who are deeply involved in the design of digital electronic control systems. As one former employee of Cincinnati Milling told the author a few years ago: "I left Cincinnati Milling because it was beginning to look more like an electronics company than a machine tool company."

Although the contracts for the Air Force's bulk-buy program were issued in 1955, more than two years elapsed before any machines were ready for shipment. This two-year interval gave the airframe manufacturers an opportunity to make their preparations for introduction of the machines to the production floor. Programming for numerical control posed a particularly serious problem: the airframe builders' programming staffs had no experience with NC, and no NC machines were available for experimental use. Furthermore, the linguistic jargon of mechanical designers was strange to programmers who had been dealing mainly in scientific and mathematical areas. The ultimate solution was the APT programming system developed jointly by the AIA and Servo Lab. Here once again we see not only the important role played by industry in the evolution of NC—this time by the aircraft in-

FIG. 6.4. Kearney and Trecker's profile milling machine equipped with a Bendix tape-control system and supplied to Northrop under the Air Materiel Command's machine tool procurement program. Courtesy of Kearney and Trecker.

dustry—but the value of the interaction between members of the academic and industrial communities. Acting alone neither group was in a position to develop APT; together they could and did.

The Servo Lab–AIA cooperative APT project relieved the machine tool builders of the responsibility for developing their own NC software. This fortuitous circumstance lessened the strain placed on the builders at a time when they were struggling to bring their products to market. The cooperative project was in contrast to computer industry practice in the 1950s whereby manufacturers of general purpose computers customarily supplied not only the hardware and software, but also applications programs.

The years 1955 through 1965 were a time of rapid expansion for numerical control. At the 1955 Machine Tool Builders' Show only one or two rudimentary point-to-point numerically controlled tools were exhibited, but numerical control was the glamour product line at the Show held in Chicago on September 20 and 21, 1965. At least 27 companies exhibited continuous-path and point-to-point NC machines for a variety of machining operations; some companies even transported potential customers to their plants to view NC machines that were too large for the exhibition hall (6–17). Elephant-

FIG. 6.5. An early Kearney and Trecker five-axis skin milling machine. Courtesy of Kearney and Trecker.

size machines of the types included in the Air Force's bulk-buy program were still an important product line, but it was evident that there was now a much broader array of tools aimed at customers in other industries. Exhibits included NC machines dedicated to drilling, boring, milling, or lathe turning, but by far the most popular machine was the so-called "machining center," a versatile tool which could perform drilling, boring *and* milling under digital command, and through use of a digitally controlled tool rack and changer, could select the appropriate cutting tool for each operation. One especially noteworthy tool, Kearney and Trecker's Milwaukee-matic, could accommodate a total of 15 tools; these could be selected automatically, either on the basis of job type or for the purpose of preserving machining accuracy by replacing a worn tool after a predetermined number of cuts. The 1965 Show also revealed that electronics companies such as General Electric, Bunker-Ramo, and Superior Electric were now in the business of supplying digital control equipment to machine tool builders. In addition, intense interest in NC was shown by European manufacturers and customers, and equally significant were the predictions being made (in 1965) about the applicability of NC in areas other than metal cutting—for example, for the manufacture of printed circuit boards and for integrated circuit fabrication—predictions that became reality in the following decade.

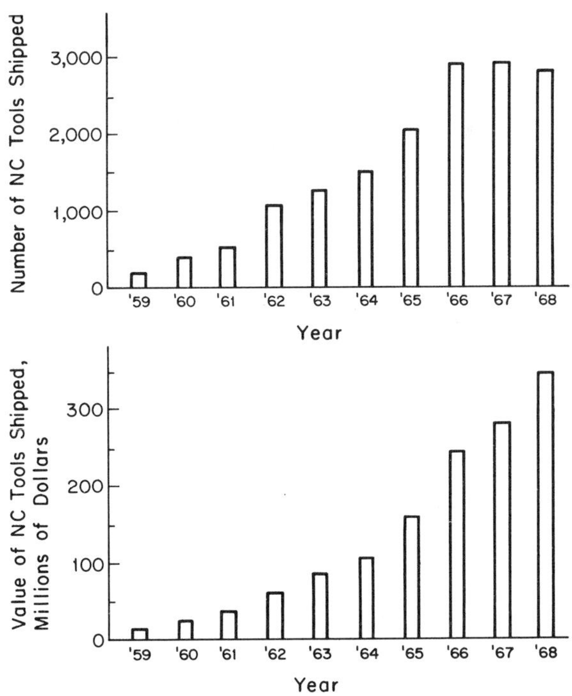

FIG. 6.6 Number and value of domestic numerically controlled machine tool shipments, 1959–1968. Source: US Department of Commerce.

Figure 6.6 shows the growth in domestic shipments and sales of numerically controlled machine metal-working tools from 1959 to 1968 (6–17). In addition to the volume of shipments and their value, the US Department of Commerce statistics for those years reveal several other interesting facts. During 1964, for example, a total of 1514 numerically controlled metal-working machines were shipped, with a value of $100 million. Point-to-point positioning machines accounted for 76 percent of the shipments; the remaining 24 percent represented continuous-path machines activated by tape, punched cards, or other forms of control. On the other hand, continuous-path machines amounted to 31 percent of the *total value*, an indication of the more complex nature of this equipment. The value of the controls for the 1514 NC machines amounted to 29 percent of the total value of the machines; controls specifically intended for continuous-path machines amounted to only 32 percent of total machine value. Thus, the initial worry, in the early stages of NC development, that the digital controls would dominate the cost of the overall system proved to be ill-founded.

Figure 6.6 also indicates that NC machine tool sales were not immune to the cyclical behavior characteristic of the machine tool business as a whole. During 1966 and 1967, NC shipments were essentially flat; in 1968, shipments turned downward slightly, in accordance with the pattern of all metal-working tools that year. Despite the cyclical nature of the business, sales of numerically controlled cutting tools continued in an upward trend in succeeding years, peaking in 1981 at a value of almost a billion and a half dollars—$1,467,000,000.

In addition to a steady growth in sales, the numerically controlled machine tool business underwent continual change over the years. Whereas in the 1960s and 1970s emphasis was on stand-alone machines, the trend in the 1980s has been in the direction of a battery of various types of NC machines controlled by means of a central computer. Parts undergoing machining are transported on pallets between machining centers (often called cells) over automatically controlled transport systems. The goal of the overall process, frequently referred to as flexible manufacturing, is an improvement in machining efficiency through speeding up the process and increasing the use factor of each machine; under computer control, a variety of parts can be routed simultaneously to the appropriate machining stations. These modern, flexible manufacturing systems, with their diverse machines, central computer control and automatic materials-transport equipment, have extended the applicability of the Detroit-automation mass production concept of the 1950s to small job-lot production. Figure 6.7 illustrates a modern flexible-manufacturing installation.

In addition to the Aerospace Industries Association, a second industry trade association, the Electronic Industries Association (EIA) also made important contributions to numerical control during the early years of its development. Through its Standards Committee, the EIA eased communication among those working in the NC field by promulgating a set of standard definitions of relevant NC terms, and also circumvented possible chaos in the realm of input-storage media by issuing a set of standards for perforated paper tape and magnetic tape used to store input-command data. Standardization in newly emerging technologies is a delicate matter of timing: on the one hand, standardizing too quickly will often stifle incentives for further innovation, while on the other hand, prolonged delays in setting standards can impede progress. In defining input-tape standards, the EIA made a timely contribution to a stage in the machining process that would benefit from standardization without inhibiting competition among system suppliers.

The Numerical Control Society

So intense was activity in numerical control in the early 1960s that representatives of user and supplier groups working in the field thought it appropriate to examine the wisdom of founding an organization or society to serve their interests. Through the initiative of William M. Stocker, Jr, senior associate editor of *American Machinist* magazine, a McGraw-Hill publication, some 35 people were invited to discuss the matter at a Founders Convention held on June 14–15, 1962 at the Manhattan Hotel in New York City. Present at this meeting were representatives of machine tool builders, numerically controlled equipment users, and control system suppliers. Also invited—wisely—was an attorney with expertise in federal anti-trust and tax law. As one might expect, when such a diverse group of relative strangers is asked to

FIG. 6.7. A numerically controlled flexible-manufacturing installation. Courtesy of the Giddings and Lewis Machine Tool Company.

pass judgment on whether or not a new organization is needed, initially there was little coherence to the discussion and much venting of positive and negative views (6–18, 6–19). For a while there were more questions raised than answers. Some thought that too many organizations already existed and that a new one was not needed; their position was that the existing professional societies representing the electrical, mechanical, machine tool, and production communities should be able to serve the numerical control field adequately. Others, citing the broad-gauge nature of numerical control, pointed out that the new technology not only cut across the traditional disciplines of mechanical, electrical and hydraulic engineering but embraced the newer fields of computers and feedback control as well. Also, it was predicted that although numerical control was then being applied principally to machine tools, the technology would eventually find its way into other forms of manufacturing. None-the-less, agreement was finally reached that, in order to adequately support such a broad cross-section of disciplines, an entirely new organization was needed.

At first it was unclear just what this new organization should be. Should it take on the attributes of a trade association? Should it be a professional society? Should it be formed as a special group to work with committees or subcommittees of existing professional societies? Or should an entirely new organization be established? After an all-day debate, consensus was finally reached that a new organization would be formed and that its tentative title, pending formal legalization, would be the Numerical Control Society. The principal objective of the Society, as finally settled upon, was the advancement of numerical control through education of its users and potential users. Areas to be covered included facilities and equipment; control systems; data processing (including computer applications); organization for numerical control; inventory effects; and performance measurements and evaluation.

The founding of the Numerical Control Society gave important further impetus to numerical control development and refinement. Through its professional journal and its annual international conferences it has provided a forum for the exchange of ideas and the recognition of accomplishments. The Society's first Joseph Marie Jacquard Memorial Award was presented to John T. Parsons in 1968. Gordon S. Brown, James O. McDonough, and William M. Pease were the 1970 recipients, and Douglas T. Ross shared the award in 1975. Membership in the Society has been as high as 4000. In keeping with all the technical advances in numerical control over the years, and its penetration into many industries, the Society recently changed its name to the Association for Integrated Manufacturing Technology.

Still further impetus was given by the Society of Manufacturing Engineers,which formed the Computer and Automated Systems Association (CASA) to respond to the NC–APT–CAD–CAM evolution.

Summary

The contributions made to numerical control by the Air Force, Servomechanisms Laboratory, and industry can be grouped into three periods: 1949 to 1953, 1954 to 1959, and 1960 to 1970. These intervals correspond to the development, transitional and expansion phases of the technology, respectively. In 1949, when the Parsons–MIT contract was the Air Force's sole NC project, the intensity of government effort was identical to that of Servo Lab. Concurrently, during the development phase, industry was involved to some extent in explorations of numerical control methods through use of its own research and development funds. In the aggregate, however, the intensity of industry effort was relatively low until the end of the development phase, about 1953. At that point, several things happened. The Air Force decided to proceed with its machine tool procurement program; Servo Lab began to transfer its NC technology to industry through the Giddings and Lewis contract, its internal educational programs, and the Air Force-sponsored industry indoctrination program; industry began gearing up to meet the market demand for NC tools under the Air Force procurement program; and the Air Force, Servo Lab and industry undertook joint development of the APT system for NC programming.

Thus, at the end of the transitional phase, about 1959, the climate was right for the start of an expansion phase, characterized by an enormous increase in NC activity in the industrial sector. The machine tool builders, aircraft companies and others now seized the initiative and began to develop and market a host of new products based on the concept of numerical control. As this group assumed a leadership position in the hardware and software aspects of numerical control, Servo Lab, still under Air Force sponsorship, withdrew from the APT development project and turned to the related area of research in computer-aided design through the use of numerical methods. This research was terminated in 1970 when once again industry assumed a leadership role in the design and fabrication of computer-aided design and computer-aided manufacturing systems.

7

THE LESSONS LEARNED

When numerical control appeared on the scene in the early 1950s, it was more than just a simple extension of the technology of automatic control. It represented a marked discontinuity in the way automatically controlled processes had been traditionally carried out. Executing real-time control through use of discontinuous (digital) signals corresponding to numbers, rather than through use of continuous signals that are analogs of physical quantities, was as dramatic a departure from the norm as was use of alternating current power in place of direct current power sixty years earlier, or the displacement of vacuum tubes by transistors in the 1960s, or the emergence of the automobile as a mode of transportation at the turn of the century. In all these examples, the new technologies opened entirely new paths, completely separate from the old ones, along which further advancements in their respective fields would be made and from which myriad unforeseeable applications would emerge. These applications, in turn, would not only require a host of new components for their implementation, but in many instances would create opportunities to look at new ways to carry out processes as a whole. Thus, in the case of numerically controlled machine tools, a family of new sensors emerged which could measure quantities such as linear and angular distance and corresponding velocities and express these quantities directly in digital form rather than as an analog signal. A market was created for a new class of digital computer—one that could be integrated with sensors, controllers, and machine tools and be dedicated solely to real-time digital control of the total system. Numerical control created an opportunity for a breakthrough in the concept of computer programming. The idea of a hierarchically structured computer language which enabled the programmer to express commands in simple, English-like words, rather than in highly cryptic, codified form, and which was able to direct the machine to compile subprograms automatically into an integrated system program, received a big boost from numerical control because there was a clear-cut need for such a computer language in this application. The need propelled the research.

Numerical control had a major impact on manufacturing and manufacturers' organizational structures. As Kenneth Wood of Boeing expressed it at an MIT lecture in 1957: ". . . it is a new manufacturing philosophy: *numbers completely controlling machines*. . . . [it] is a manufacturing system requiring a new manufacturing recipe."(7–1) Traditionally in the aircraft business, a mechanical part, hogged out of a block of metal, was the result of a sequence of steps and operations carried out by various departments operating in relative isolation. To achieve the full potential of NC, a different structural organization was required—one in which there was much more interaction among all who were involved in the whole process from design to production. The new manufacturing recipe called for a blending of the skills of planners, tool engineers, machinists, programmers, and the design engineers themselves, all acting in concert. Each had something to contribute as the format of the complete manufacturing process evolved. Fortunately, the aircraft manufacturers, the first group to incorporate NC machines into their facilities, saw the importance of an integrated team approach very early, and were thus able, through careful planning, to take full advantage of NC as the new machines reached their production floors. Their trade association, the Aircraft Industries Association, through its Subcommittee on Numerical Control, kept its membership informed about progress being made during the development of NC and what organizational changes should be made in order to take full advantage of the new technology.

What can happen when management fails to prepare adequately for new technology is highlighted in an article by V. Petrov, Senior Setup Man of the Division of Program Controls of the Scientific Research Institute of Machine Building Technology, of the USSR. In his 1962 article (7–2), Petrov pointed out that in Leningrad over two hundred machine tools with program controls remained virtually idle at the time of his report. He complained that leaders of the administration and plant managers "were not rushing into this important matter," although they were not wholly to blame for the inaction: plants were not supplied with the basic electric diagrams of the machine tools nor with the operating and maintenance manuals. He then cited the need to "organize without further delay the training of lathe operators, maintenance machinists, and electricians at plants possessing machine tools with program controls."

In 1949, numerical control was a technology whose time had come. This fact had a bearing on the speed at which NC could be brought to fruition. The challenges resided in configuring the digital control *system*, not in the development of components; system research could go on uninhibited by constraints imposed by components. The principles of electronic digital computers and analog servomechanisms were well documented in 1949, as was the theory of digital pulse circuit design, all of which had reached a highly refined state of development through the enormous effort put forth in these areas during World War II. If there was any uncertainty in the technology itself, it was in the practical implementation of *digital* servomechanisms, the theory of which was a postwar achievement. Thus, we conclude that the idea of merging digital computation and digital servos into a real-time control system was indeed timely. Had it come any sooner, it is possible that the whole philosophy of digital control might well have languished.

However, much more had to be done than just putting together a digital data processor and servo system that would accept data and control machine tool motions. One way to do the job would be to use the new technology to automate what a manual operator does as he manipulates the controls of his machine tool.. More often than not, however, merely mechanizing manual operations will prove to be less than ideal and will sometimes doom the automated version to failure. Fortunately, Servo Lab staff understood the potential pitfalls of simplistic mechanization and so rejected the original idea of contouring three-dimensional surfaces by drilling myriad holes to prescribed depths and smoothing the resultant surface to specification by grinding or filing. Instead, NC started at the Laboratory with digital data processors, digital servos, and a three-axis machine tool as the "givens," and then proceeded to an optimum *system* configuration that would take full advantage of the potential of each subsystem. Taking a fresh, unbiased system analysis approach paid off, as it so often does in situations that call for engineering innovation.

Although NC's time had arrived technologically, it was ahead of its time in terms of its immediate acceptance as a marketable product. The idea of controlling a process by numbers is as old as Jacquard's loom, but exploitation of it for machine tool control was essentially nonexistent in the late 1940s. Research groups of various industrial organizations were experimenting with control by numbers, but their efforts were largely regarded by management as being in the blue sky category. Thus, lack of management vision led to research budgets that were woefully inadequate for the magnitude of the effort required. As a result, when the MIT NC milling machine became a reality, skepticism about its worth had to be dispelled and a market had to be created for NC machines. Involved here is the oft- debated question of how to carry out *technology transfer* most expeditiously and effectively: who should do it and how much should be done. It would appear that the best choice of an entity to undertake the follow-up tasks is the group that did the original development. One can argue, for example, that the original developers are in the best position to articulate the capabilities and limitations of the device or system they created because they thoroughly believe in the results of their efforts and would like nothing better than to see their creation perpetuated. However, in an industrial setting, it is not always possible for an R&D group to assume the roles of the product engineer, the manufacturing supervisor, and the marketeer. There are often barriers to this mode of operation, not the least of which is a highly compartmentalized organizational structure that favors passing along the product but not its originators. Also, there is often the managerial belief that marketing types make the best marketeers, researchers are at their best when they are doing research, and each group should stick to its own specialty. Then, too, researchers by their training and professional attitudes generally prefer to remain in research. The reasons for this preference are not axiomatic but may be attributed in part to a status symbol and in part to the fact that many R&D people are just not mentally suited for roles in manufacturing and production.

The subject of new technology transfer is even more complex when the technology originates in a university setting. Universities are commonly thought of as wellsprings of new and radically different theories and ideas and as places where such theories and ideas can be developed and tested in an atmosphere uninfluenced by pressures of productivity and profit. The business of a university is education, and in the academic research laboratory the educational process is carried out by faculty and students working together on their research problems. Research results are transferred outside the academic research environment through reports, theses and other publications, conferences, faculty contacts with industry and government, and (rarely) through joint projects between a laboratory and the industrial sector. Hence, upon completion of the demonstration numerically controlled milling machine in 1952, the Servomechanisms Laboratory had a choice to make: it could write a final report on the project and move on to other things, or it could stay with the project and help reduce the new technology to practice. Withdrawal at this point would have been perceived by the academic community as perfectly normal, but there was never any serious thought given to this possibility. The Laboratory was staffed by very bright young pioneering spirits who were eager to see the results of their handiwork put into practice. They were thoroughly convinced they had created something that would someday have a profound impact on manufacturing methodology, provided they could get the technology transferred from the Laboratory to those in a position to exploit it commercially. What would have happened to numerical control had Servo Lab withdrawn as soon as it had proven its feasibility is purely conjectural. However, it is the opinion of Eugene Merchant, formerly Director of Research Planning at Cincinnati Milacron, Inc. (7–3), that: "There seems to be quite general consensus in the US manufacturing industry that, as a result of this [NC] Federally supported program, this country gained five to ten years' lead time, initially, in the innovation of numerical control technology in industry."

Servo Lab's decision to continue the project beyond the proof-of-feasibility stage was one step along the path to general acceptance, but a bigger step was the plan it designed in order to ensure NC's success. The principal ingredient was *education*—education of machine tool users, suppliers, and the electronic and mechanical designers of the machines themselves, with equal attention paid to all three groups. This formula certainly was not a new one in the field of engineering, but in light of early indifference to the NC concept, an extraordinarily intense effort on all three fronts was mandated, the lack of which could have led to protracted delay in the acceptance of the new idea, or even failure to put it across. In retrospect it was this early concentration of the Laboratory staff on the end users—members of the aircraft industry—that proved to be a good move. Even though continuous path numerical control was ideally suited to the small-lot production requirements of the industry, it could really have no impact until this user community recognized its value and created a demand. The machine tool builders would be in no hurry to move into a technology for which there was no market.

Hence, the Laboratory's aircraft industry indoctrination program, consisting mainly of plant visits and demonstrations at the Laboratory, accomplished two things: it created an awareness at various management levels of the potential of this new technology, and in so doing it laid the foundation for a future market for the technology. The indoctrination of the aircraft industry's and machine tool builders' engineers was materially enhanced through the three intensive educational courses offered at MIT in 1954, 1955 and 1957.

Although we concentrated our indoctrination efforts on the aircraft industry, our plant visits and Laboratory demonstration programs also included the machine tool builders. Once again this plan paid off, since one company, Giddings and Lewis, decided to go into NC in a serious way, as described in Chapter 4. Our continued presence in the NC field undoubtedly paid off indirectly too; soon after our public demonstration, various machine tool builders stepped up their own in-house NC research and development programs along lines that best suited their ongoing organizational styles.

The time span between the conception of an innovative idea and its implementation in the marketplace is often longer than innovators think it should be, undoubtedly because of the convictions and enthusiasms of the innovators. Two time components are involved: the time required to achieve demonstrable results (often longer than the innovators originally estimated, or than impatient managers would prefer) and the time needed for the product or process to gain acceptance in the marketplace. On the latter point, a *sound* innovative technology is essential. Even though potentially useful *elements* of technology are at hand, *the way in which these elements are synthesized into a system is crucial.* It is not possible to sell an empty box—no matter how pretty the package or how glowing the product description, its emptiness eventually will be discovered. The time required to establish numerical control as an acceptable machining process was shortened because of several factors: our funding was always at or above critical size, that is, the financial resources at our disposal allowed us to work near the limit of our capabilities, and fortunately our research and development staff was eager to undertake an industry indoctrination program in order to establish the viability of the process. Most importantly, the Air Force came along at the right moment in the development phase with its numerically controlled machine tool procurement program, thus ensuring a market for the product. In addition, the exposure factor was high throughout the project; the research results were intentionally made highly visible.

Technological innovations sometimes create unforeseen problems that did not exist in the older order, and complete fulfillment of new objectives may depend upon solutions to those problems. In the horse-and-buggy days, a simple utterance of "giddap" and a flick of the buggy whip were all that were needed to get one's transportation started. When the first automobiles came along, however, the job was not that simple, as anyone knows who has ever had to hand-crank a balky engine on a cold day. Eventually, the invention of the self-starter motor solved the problem, and a gentle push on the starter button became the electrical equivalent of a vocal "giddap." So it was with numerical control. Initially the automaticity of the computer-controlled machine tool came at the added expense of having to hand program the computer-controller. Another development was needed—the APT system of programming—before the overall numerically controlled machining process could be fully justified economically. Later developments in related fields, such as transistor and integrated circuit technology, contributed further to the attractiveness of the original innovation. The "unforeseeables" inevitably add to both the cost of an innovation and the time required for it to gain popular acceptance.

During the first ten years of the project, the Air Force invested 1.05 million dollars in Servo Lab for the development of the experimental numerically controlled milling machine, the APT project, and the information-dissemination program; another 3.0 million dollars was invested over the second decade for research on computer-aided design and manufacturing. In light of these expenditures, a valid question to ask is: What was the government's return on its investment? This question strikes at the heart of a more fundamental issue: how one should measure the return on public funds invested in research. Should it be in kind only? If so, how should the dollar return be calculated? Alternatively, if financial return is not the proper yardstick, or only partly so but essentially uncalculable, what other criteria can be applied to test the worth of the investment? In responding to these questions, we shall comment first that some broad, qualitative statements can indeed be made about the monetary return on government-supported research, but we shall defer detailed commentary on the financial issue until we cite other kinds of returns that should also be considered. In the case of MIT's numerical control project, the Air Force's motive in sponsoring the project was national preparedness. It wanted to ensure that its metal-cutting capability would not be a bottleneck as it moved its high-performance aircraft program forward. Productivity improvement in metal-cutting was a second goal, and assurance of a modern, healthy machine tool manufacturing base was a third. Measured against these objectives, it is clear in retrospect that the project yielded an ample return on investment. At the end of the first ten years there had emerged an entirely new method of controlling continuous-path metal-cutting machines automatically, and a new special purpose computer language that ensured the cost-effectiveness of the process. During the second ten years, foundations were laid for computer-oriented manufacturing as a whole.

In addition, there were other beneficial fallouts from the project. The machine tool industry was brought into the modern computer age, thereby strengthening the country's manufacturing base not only for national preparedness purposes but for the common good. Similarly, there was created an awareness of the potential of computer-aided design and computer-aided manufacturing among users and suppliers of those kinds of systems, and tangible contributions were made toward their practicality. Measured in these narrow terms alone, the Air Force accomplished what it set out to do and in this sense got a favorable return on its investment.

However, there were additional returns on the investment, some of which can be grouped under the heading of "the amplification factor" or, as expressed in a Bruce Olds Associates Inc. report: "the chain reaction effect" (7–4).

From this viewpoint, the numerical control project involved *people* and *new technology*. In terms of the people component, a few senior Laboratory staff members were able to provide on-the-job education in forefront technology to many, many others, all of whom eventually took up positions in industry or education where they in turn could lead and guide others along new paths of endeavor. With respect to the second component, the new technology generated in the Laboratory found its way into industry through the technology transfer process, and into educational institutions where it was imparted as updated, modern education.

A chain reaction or amplification effect was touched off in another direction. As knowledge of numerical control methodologies proliferated, it was inevitable that the new technology would find its way into industries outside the metal-cutting field. Over the past twenty years, numerical control and information processing by numerical means have largely superseded analog techniques. In the narrow terms of the Air Force's interest, the advent of digital controllers as replacements for analog controllers opened up vistas of opportunity for improved accuracy and reliability in advanced military systems for navigation, guidance and fire control, and in the non-military world, to improvement in digital control of processes in general.

Since this book centers on feedback control technology, it is appropriate to end it by showing diagrammatically how it is possible for the government to receive a favorable financial return on its investment in engineering research conducted in an educational environment. Figure 7.1 shows that the

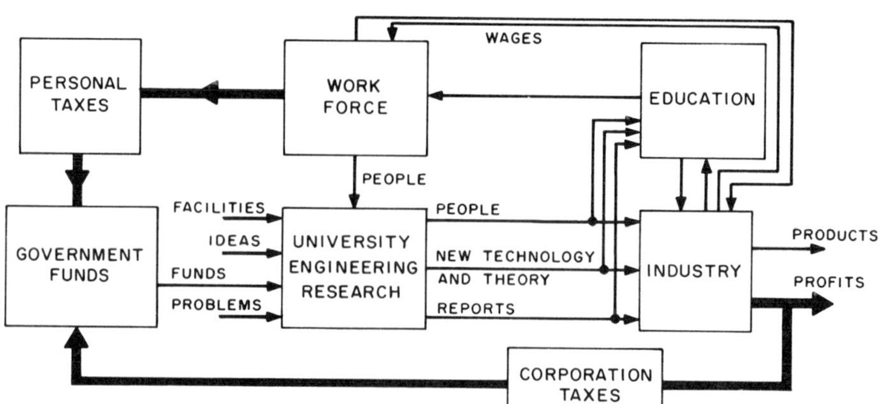

FIG. 7.1. Interrelationships among the government, university, and industry, with emphasis on the Government's return on its research.

engineering research laboratory inputs are problems, people, ideas, facilities, and government funds. The outputs are research results (new theory and new technology), reports and people—people being the students and staff who do the research and then leave the laboratory. Laboratory outputs go into industry and education; the outputs of education are people who find their way into industry or back into the university research environment. Industry's outputs are products and profits derived from the sale of products. As shown by the heavy lines in Fig. 7.1, the feedback (or return on the original investment) is derived from taxes paid by the corporation on profits from products involving the new technology, and from personal income taxes paid by the employees engaged in the manufacture of new products. Most people will agree that the financial amplification factor, as shown by the *heavy* feedback lines in the figure, in contrast to the *light* line for input funds, was a handsome number in the case of numerical control, even though its exact magnitude may never be tallied.

AFTERWORD

The emergence of numerical control put a major discontinuity into the US machine-tool industry's traditional product line. Through this single invention the industry was thrust into the age of digital electronic machine-tool control, and opportunities were created for an assortment of new products which could be offered in the world marketplace. To take full advantage of these opportunities, however, the machine-tool builders had not only to accept the new technology but also to expand and exploit it vigorously on a sustained basis. Progress in this direction was slow because of initial skepticism about cost and reliability, but eventually apathy gave way to enthusiasm and for a time the industry became a world leader in the manufacture of numerically controlled machine tools. As time went on, however, the industry failed to hold its premier position, so that by 1988, it found itself lagging in terms of US and world market share. The questions are: What happened? What went wrong? And what must be done to restore the business to a position of world leadership?

To answer these questions it is first necessary to understand the nature of the machine-tool business. It is inherently cyclical, principally because machine tools last a long time. With care, maintenance, and occasional reconditioning they can perform well for many years. Thus, during austere times the tendency of users is to "make do" with what they have and to defer new purchases until prosperity returns, or until new tools are required for a specific application or to replace worn-out ones. Between 1925 and 1950, about 50 percent, on average, of US metal-cutting tools on users' floors were at least 10 years old; the number climbed temporarily to 67 percent at the start of World War II; in 1949 16 percent of the metal-cutting tools in users' inventories were over 20 years old (A–1).

In the twentieth century the cyclicality of the business was exacerbated by the Great Depression, which was sandwiched between two major wars. During World War I the machine-tool builders responded to the enormous demand for their products, only to be faced with huge surpluses immediately thereafter. There was an upturn in the late 1920s, but the 1930s were lean years as the country struggled to work its way out of the depression. Then, in the 1940s the up–down scenario repeated—boom years during the war, slack times afterward. As an illustration, the industry shipped 99,000 cutting tools in 1919, but only 21,000 in 1921. In 1932 the number was a mere 5500, but it expanded to 307,000 in 1942. Then, after the war the number fell again to 34,500 (A–2). Many companies had to take substantial losses during some of the depression years.

In light of these violent business swings, owners and managers took a conservative position in order to stay in business. Inventory was maintained at a low level to preserve working capital. Furthermore, research and development, what there was of it, was conducted more for the purpose of improving the precision, durability, and productivity of traditional products than for the exploitation of imaginative ideas of the creative minds in the industry. Emphasis was on developments that solved immediate problems, not on major inventions nor innovative research.

It was these owners and managers, steeped in conservatism, who had to cope with the strange new technology of NC. They were surrounded by engineers, machine designers, and technicians with excellent skills in the mechanical domain, but engineers with expertise in NC technology—digital electronics and digital feedback control—were essentially nonexistent in their organizations. Thus, management found it relatively easy to find excuses for ignoring the new technology. At the Laboratory's demonstration of its experimental NC system (see Fig. 2.2), it was evident that the attendees were worried about equipment cost, programming, and reliability. In light of these major concerns and the fact that no ready market for NC tools was visible, the industry saw no reason to brashly embrace the new technology.

The information-dissemination program of the MIT Servomechanisms Laboratory and the government's Machine Tool Modernization, Selective Augmentation, and Replacement Program did much to develop a market for NC machine tools and thus provide an incentive to take the technology seriously. Since both programs were addressed to the aircraft companies, the aircraft industry, including its suppliers, became the machine-tool builders' first NC market. Although this industry was only one of many potential markets for NC tools, it was a big market when measured in terms of the size of the tools, their unit cost, and the numbers purchased. These "elephant tools," as they were called, represented a substantial income to those who could build them. On the other hand, only a handful of machine-tool companies, "the biggies," were able to handle this kind of business; the small builders were unable to compete.

Even the big companies had a hard job finding competent electronic engineers who could design real-time digital controllers. Hence, rather than building an internal staff, most companies initially chose to join forces with electronically oriented companies that already had some capability in the digital domain. Thus, organizations such as General Electric, Bendix, Bunker-Ramo and others were drawn into the numerical control business as suppliers of digital-electronic controllers. This path had its weaknesses, however, since each tool builder and its electronics supplier had to educate itself in the other's business and to develop smooth intercompany relationships. The concept of a project engineer who could take charge and direct the development of the overall tool system, including the machine itself and its controls, was less well understood in the machine-tool industry than in the electronics domain. Even when understood, hiring highly qualified project engineers was difficult because of their short supply.

In retrospect it was fortuitous that the aircraft industry provided a market for the NC tools in the late 1950s, for other manufacturing industries failed to see the value of NC until its worth had been proven on aircraft production floors. Even the auto industry was slow to discover how NC could be utilized in its manufacturing operations. Because NC enabled design changes to be implemented quickly, it eventually proved to be highly effective in the fabrication of complex experimental and prototype mechanical items such as engine blocks; however, several years passed before NC found its niche in auto manufacturing.

Although the first NC machines were the elephant tools ordered by the Air Force for the aircraft industry, an assortment of small and large US-built NC tools appeared on the market in the 1960s. Large and small companies alike became producers. Their products included discrete-positioning machines for automatic drilling operations, NC lathes and versatile NC machining centers that could do a variety of metal-cutting tasks with one stand-alone unit. The years between 1960 and the mid-1970s were upbeat years for US manufacturers of numerically controlled machine tools.

Toward the end of the 1970s, however, the US machine tool builders lost their hold on the US market, and in the early 1980s severe setbacks were encountered. Between 1980 and 1983 orders dropped by nearly 80 percent (A–3). This time, however, the decline could not be blamed on a downturn in the US economic cycle. The US economy began to expand in 1983 but US manufacturers chose to buy a substantial percentage of their new tools from foreign suppliers This was a new twist, for the US tool builders had always considered the US market to be exclusively theirs and their competition to be other domestic machine-tool builders. Even though the machine tool business showed signs of recovery in 1988 from its early 1980s doldrums, prosperity remained elusive.

The decline of the machine tool industry in the 1980s has been the subject of much study, as researchers try to pin down underlying causes. Opinions are many, and feelings about what happened run strong and deep. In one scenario, the root cause of the industry's problems in the 1980s is attributed to the MIT Servomechanisms Laboratory's numerical control program and to the Air Force, its research sponsor in the 1950s. According to this scenario, MIT's approach to numerical control was needlessly complicated both with respect to the equipment it developed and the programming required to operate it. This complexity led the majority of users and suppliers to conclude that the new technology was a laboratory curiosity that would be unmarketable; it was deemed too complicated, unreliable, and costly. As a result, the industry delayed too long in exploiting the new technology.

David F. Noble, in his historical account of Servo Lab's development (A–4), puts great emphasis on the complexity issue. He concluded that the Laboratory's demonstration NC machine tool and its APT programming system were needlessly complicated and expensive. Noble contends, in essence, that implementation of the point-to-point tool-positioning system originally set forth by the Parsons Corporation would have led to much simpler, more reliable, and less costly equipment than MIT's incremental-coordinate continuous-path metal-removal concept, and had the Parsons concept been exploited the industry would not have been misled down a path of needlessly complicated NC tools. Furthermore, had the Parsons approach been adopted, Noble's argument goes, the highly complex APT system of programming NC tools would not have been needed.

Meldman (A–3) indicates that the Air Force unwittingly aided and abetted the delay in the application of NC to small, high-volume machine tools because the Air Force's bulk buy program favored only a few large US builders who were capable of building the huge NC machines demanded by the aircraft industry. These large companies, flush with orders, strapped for qualified in-house personnel, and preoccupied by the job of making deliveries on NC tool systems with which they had little background experience, delayed too long in applying basic NC technology to a broad spectrum of new products. Meanwhile, the early emphasis on big NC machines by the large companies created a mindset within the industry that discouraged others from applying the technology to small, simple machines. When the bulk-buy program was finally over and the builders became free to turn their attention to other types of tools, it was too late; foreign competitors had already seen a broad NC market opportunity and were able to gain a foothold in that market.

Meldman also states that the close relationship between the NC machine tool builders and the military in the 1950s and 1960s lulled the builders into a state of complacency on matters pertaining to competition and productivity. Says Meldman: "The firms engaged in this [NC] effort found that the cost of the equipment mattered less than is capabilities, since the Air Force was paying for developing the tools and soon became the most important customer" (A–5). Meldman further comments that tied to this argument was "the Department of Defense policy in the early 1960s whereby costs [of weapons systems] were allowed to increase at a steady rate based on past increases, rather than being tied to actual engineering estimates." Thus, contracting firms had strong incentives to run up costs.

The underlying questions with respect to the above scenario are these: (1) Was Servo Lab's approach to numerical control unnecessarily complex; and (2) Is there a clear-cut linkage between MIT's numerically controlled machine tool technology of the 1950s and the decline of the industry in the 1980s.

To answer these questions it is important to bear in mind why the Air Force decided to sponsor work in the realm of automatic machine tool controls. As explained in Chapter 6, after World War II there was doubt in the minds of Air Force officials, the machine-tool builders, and the airframe manufacturers about the ability of conventional tool-control methods [template control, for example] to meet machining accuracy requirements for the advanced military aircraft being planned. The Air Force therefore sought new technology that could meet its future machining requirements, and it was willing to explore highly unconventional ideas that might meet its goals On the other hand, the Manufacturing Methods Branch at Wright-Paterson Air Force Base was development-oriented; it was not a basic research sponsoring agency. It was looking for unconventional metal-removal methods based on the latest available technology that could be put to practical use on aircraft production floors within a few years, not in the indefinite future. This time constraint is important because it shaped the thinking of Servo Lab staff members and the approach they ultimately took to the development of their real-time digital control system. They chose not to demonstrate merely that digital control technology would work in principle; on the contrary their goal from the outset was to develop an authentic automatic computer-controlled machine-tool system that was ready to sculpture representative aircraft parts at cutting speeds comparable to those of conventional machines. Their purpose was to prove that the technology was ready for practical use and that it would be unnecessary to wait for future advances in relevant technologies and components. And this they did.

This issue raises the question of what complexity really means in the context of technology. Some equate complexity to size or to costliness, whereas others reserve the term for an assemblage of parts or components that are interconnected in a highly intricate way. Others think of complexity whenever there is an abundance of interconnected components, regardless of the simplicity or intricacy of their interconnections. Complexity is often confused with strangeness, especially when a device or system is highly revolutionary and the first of its kind. The ease of mastering the device or system may then determine whether the device or system is deemed complex or merely strange. To the developers of new technology, the technology is never strange nor complex; it is commonplace from the beginning. To others, new technology may appear strange and complex at first, but as underlying concepts are grasped, these labels quickly vanish. To those who opt not to learn and understand the technology, it is likely always to appear complex and unfathomable.

Whether or not Servo Lab's incremental-coordinate, continuous-path machine-tool system was an unnecessarily complex contraption is really a nonissue. There was no viable alternative. The Parsons Corporation's Cardamatic approach was a possibility, but a commercially practicable Cardamatic machine was physically unrealizable at the time. The application the Parsons Corporation had in mind when it sought help from Servo Lab was the contouring of three-dimensional parts like those found on aircraft. Since many complex shapes were involved, the application called for a continuous-path metal-removal process. However, the Cardamatic concept was a point-by-point, discrete-position cutting process. Furthermore, there were technological barriers which prevented a commercially acceptable implementation of the Cardamatic approach. Cutting tool position information could be stored on punched cards but the cards could not be read fast enough to satisfy required cutting speeds. Similarly, a digital computer could have been used to read out information at required rates, but the amount of digital memory required to store the information involved in machining representative aircraft parts on a point-by-point basis was beyond the state of computer-storage technology in 1950. It was therefore necessary to search for another system configuration that could compete successfully with conventional machine-tool capabilities. The incremental-coordinate continuous-path system that was invented did, in fact, meet commercial requirements both in terms of cutting speed and the class of parts to be machined.

It has been asserted that the APT system of programming machine tools was also unnecessarily complex and therefore set the machine tool industry back, rather than forward. APT is indeed a big programming system that took many man-years to develop. However, once it was developed it was an extremely easy system to use. Its English-like inputting vocabulary could be easily mastered by skilled tradesmen—it was designed that way—and furthermore it was *open-ended* and could be easily expanded, as needed, without having to scrap it and start all over. APT has withstood the test of time; it is still a world standard special-language programming system.

When APT is labeled unnecessarily complex, one must provide a reference against which complexity is measured. The predecessor development in Servo Lab, the only alternative at the time, was based on the concept of storing, in a computer, a library of subroutines that could be called upon to meet the needs of the programming job at hand; but the subroutine concept had drawbacks. As time passed, a tremendous number of subroutines would be needed in order to meet every foreseeable and unforeseeable need, and the use factor of any one routine would likely be low. A low use factor in turn equated to an inefficient use of a large amount of computer storage—a precious commodity in those days. Thus, APT was a solution to the programming problem, not an impediment.

With respect to size and complexity, there is no theorem which states that within a problem class one should tackle the simple, trivial case first, then work upward toward more complicated cases. Individual circumstances govern the approach. Sometimes the general case is not immediately identifiable; it evolves only after experience is gained by solving simple examples. Other times it is the style of the researcher that governs the approach: some like to work from the general to the particular, others prefer to start at the bottom and work upward. Occasionally, researchers have no choice: external conditions dictate the starting point. That was the case when the numerical control problem was brought to Servo Lab. The Laboratory received a contract to investigate a digitally controlled machine tool system that could hog out complex-shaped three-dimensional parts from billets of metal. It had neither a charter nor funds to work on, for example, a digitally controlled drilling machine, a much simpler problem. Such a charter would undoubtedly have led to a much simpler control system, but numerically controlled drilling machines would not have solved the three-dimensional contouring problem, which was the problem in which the Air Force was interested.

There is a deeper issue behind the development procedure employed at Servo Lab. It involves the matter of complexity when attempts are made merely to automate a manual process by mechanizing each step of the manual process on a one-for-one basis, as the Cardamatic approach proposed to do. Often, such an approach leads to a dead end or to a more cumbersome configuration than would have resulted had a fresh, new systems approach been taken. In the systems approach, the one taken by Servo Lab staff, the manual steps in the process to be automated are ignored. Instead, the automation process starts from scratch with a thorough systems study that takes into account basic performance requirements and technologies that show promise of meeting the desired performance. In the systems approach many of the resultant steps in the automatic system bear little resemblance to those of its manual counterpart, and performance requirements can often be met with a simpler configuration.

Any link between the decline of the US machine tool industry in the 1980s and the original numerical control technology itself is far-fetched. Other factors, some under management's control, others not, conspired to dislodge the US builders from the world-leadership position they enjoyed prior to the recent decline. As early as 1955, various companies in the industry did, in fact, have numerical control research programs under way, and these programs lead to a variety of new products, including large and small continuous-path and point-to-point NC machines for drilling, boring, milling and combinations of these operations. As pointed out in Chap. 6, domestic sales of NC machines reached a total of $ 1.467 billion in 1981, a banner year for the industry.

However in the mid-1950s US owners and managers were exploiting NC technology mainly to gain a quick advantage over their domestic competitors; they were not looking ahead to the day when their major competitors might be off-shore manufacturers. In 1955 for example, MIT's Servomechanisms Laboratory submitted to the National Machine Tool Builders Association (NMTBA) an outline of projects the Laboratory was interested in undertaking jointly with one or more machine tool companies, either directly or under auspices of the NMTBA. Included in the list were projects relating to digital controls for discrete positioning machines that perform drilling, boring, and riveting (A–6). Had these projects been implemented, they would have substantially broadened the base of the industry's NC product line, particularly with respect to tools for small shops. The proposal was turned down, however, on grounds that: "most machine tool companies with an interest in numerical control have research departments of their own, and are already spending a good deal of money in this field. They prefer to keep their activity confidential and by doing so they are hoping to be able to provide a better product than their competitors" (A–7). The competitors were perceived, of course, to be US competitors.

Although the US builders hoped to meet competition through new products based on NC, they were relatively insensitive to the value of making productivity improvements through modernization of their own plants and other cost cutting measures (A–3). Such measures were not put in place soon enough to ward off inroads made by foreign builders who introduced their machinery into the US market at attractive prices and with short delivery times. Also, unfavorable international currency-exchange rates that prevailed during much of the 1980s exacerbated the US cost-competitiveness problem.

Dertouzos, Lester, and Solow in their study of the current decline of American manufacturing in the 1980s found a general malaise in US manufacturing; the consumer electronics industry, for example, has been essentially conceded to off-shore manufacturers, and the steel industry is facing stiff competition from abroad (A–8). This general diminution in the intensity of American manufacturing has pressed back onto US machine tool manufacturers, since this industry supports most other kinds of manufacturing. Hence, Dertouzos et al. attribute the decline in the US machine tool business not only to the builders' policies but to the buyers and users of tools as well. The Dertouzos study also cites, as negative influences on the machine-tool and other US industries, a weaker demand for innovation in the realm of new process technologies and the absence of a strong government policy favoring and promoting US industry.

Differences in management style among major corporations of the free world have also affected American manufacturing adversely over the past decade or so. In the United States there is a corporate culture that favors managers who turn in steadily increasing profits from year to year, and often on a quarterly basis. This culture, which some say is driven by the investment community, inevitably influences management decisions and leads to policies that maximize short term gains at the expense of longer term research projects which might eventually pay off in the form of new products or improvements in production efficiency. It is this culture that is trying to compete against others that are satisfied with less short term gain and low profit margins, but place great emphasis on production efficiency and on bringing a steady flow of new products to the marketplace.

Although the US machine-tool industry is relatively small, it is crucial to our national welfare. Because machine tools are an important part of many manufacturing operations, they are essential for a healthy US manufacturing base. Total dependence on foreign builders in times of national crisis would be calamitous. However, to restore our domestic tool production to a world leadership position will require a national resolve that includes a coordinated effort among business, government, and universities. The government can play an important pump-priming role through actions favorable to the industry. These actions might include tax incentives for machine-tool companies that modernize and automate their own manufacturing facilities and direct or guaranteed federal loans to cash-poor tool companies for plant modernization and automation; permitting domestic machine-tool builders to conduct intercompany research programs without risk of running afoul of antitrust regulations and offering tax breaks or matching funds for such programs; and providing long-term financial support for professional and vocational educational programs for managers, engineers, and skilled tradesmen in the machine-tool industry. The government can also assist by supplying financial support for research and development projects carried out jointly by universities and industry. The assumption, of course, is that the machine-tool companies have changed their 1955 negative viewpoint about sponsored machine-tool research at universities and now see the value of participating in such research. Government-sponsored research at the universities can yield a twofold return on the investments: beneficial technical fallouts to the industrial sector and the education of talented young engineers with knowledge and some research experience in the machine-tool field.

Essential to the return of the industry to a place of worldwide prominence will be the ability of the builders to project an image that will attract bright, young managers and engineers who perceive the industry as dynamic and exciting, and a place where there are intellectual challenges and opportunities to apply the latest skills of their professions. Extraordinary efforts by the industry, government, and technical universities, working collaboratively, will be required to make this happen.

That the government can have a salutary influence on the course of technology has been amply demonstrated by the Air Force's numerical control project. This government entity sponsored the research that spawned NC and it created a market for it at a critical time through its Machine-tool Modernization, Selective Augmentation and Replacement Program. But the Air Force's NC program was mission-oriented; it applied only to machine tools that would enchance the effectiveness of the government's military-aircraft manufacturing facilities. At the time there was no comparable government agency with a charter to support machine-tool development programs that would enchance the effectiveness of the nation's civilian manufacturing base, where the bulk of the wealth-creating manufacturing resides. Nor does such an agency exist today. This is in contrast to policy of other industrialized nations, notably Japan, where manufacturers of civilian goods receive government incentives to create a steady stream of new products that can compete in the world marketplace on the basis of cost, quality and short delivery times. Thus US government officials and managers of American manufacturing industries now face a challenge to devise a long-range plan that will enable the nation to meet the competition of its trading partners and eliminate the yearly international trade deficit now being experienced.

REFERENCES

Chapter 1

1–1. Redmond, Kent C. and Thomas M. Smith. *Project Whirlwind: A Case History of a Pioneering Computer.* Bedford, MA: Digital Press, 1985.

1–2. Wildes, Karl L. and Nilo A. Lindgren. *A Century of Electrical Engineering and Computer Science at MIT, 1882–1982.* Cambridge, MA: The MIT Press, 1985.

Chapter 2

2–1. Linvill, William Kirby. "Analysis and Design of a Sampled-Data Control System," MIT doctoral thesis, 1949.

2–2. Society of Manufacturing Engineers. *Manufacturing Engineering*, **88** (1): 127, 1982.

2–3. Parsons Corporation. Description of the Parsons Automatic Milling Machine project, Traverse City, MI, January 11, 1951. MIT Servomechanisms Laboratory files.

2–4. Everett, Robert R. Memorandum to Jay W. Forrester on the Cardamatic milling machine, July 12, 1949. MIT Servomechanisms Laboratory files.

2–5. Pease, William M. Memorandum to Gordon S. Brown on the conference held at IBM Corporation, Endicott, NY, July 19, 1949. MIT Servomechanisms Laboratory files.

2–6. Memorandum of Agreement between the Massachusetts Institute of Technology and the Parsons Corporation, July 1949.

2–7. McDonough, James O. Notebook entries made at a meeting with John Parsons at Servo Lab, October 24–26, 1949. MIT Servomechanisms Laboratory files.

2–8. Parsons, John T. Letter to Gordon S. Brown in follow-up to October 24–26, 1949 meeting of Parsons Corporation, MIT and the Air Force. November 2, 1949. MIT Servomechanisms Laboratory files.

2–9. Forrester, Jay W. Memorandum to Gordon S. Brown on a review of the milling machine project. December 30, 1949. MIT Servomechanisms Laboratory files.

2–10. MIT Servomechanisms Laboratory Engineering Report No. 2 (DIC 6694): "Provisional Specifications for the Development of a Numerically Controlled Milling Machine," June 30, 1950.

2–11. Frost, H. Bonnell. MIT Project Whirlwind Report R–178: "Vacuum Tube Life Experiences," 1950. MIT Libraries, Archives and Special Collections, Cambridge, MA 02139.

2–12. Pease, William M. Memorandum to Gordon S. Brown on Pease's trip to Wright Field, January 10- 12, 1951. MIT Servomechanisms Laboratory files.

2–13. Parsons Corporation. Letter to Nathaniel McL. Sage on the Termination of the MIT–Parsons Memorandum of Agreement, January 23, 1951. MIT Servomechanisms Laboratory files.

2–14. Carter, Wilbur R., Colonel, USAF (ret.) An interview with J. F. Reintjes, Rolling Hills, CA, November 14, 1969.
2–15. Webster, William M. An interview with J. F. Reintjes, Dayton, OH, October 2, 1969.
2–16. Sage, Nathaniel McL. Memorandum to Paul V. Cusick, William M. Pease and Albert F. Sise on the negotiations for a direct contract with the Air Force, February 6, 1951. MIT Servomechanisms Laboratory files.
2–17. McDonough, James O., et al. MIT Servomechanisms Laboratory final report on "Construction and Initial Operation of a Numerically Controlled Milling Machine," Part 1: July 30, 1952, and Part 2: May 31, 1953.
2–18. Verbatim transcript of panel discussions held at the MIT demonstration of a numerically controlled milling machine, Cambridge, MA, September 15–17, 1952. MIT Servomechanisms Laboratory files.

Chapter 3
3–1. Hamming, R. W. "Error-Detecting and Correcting Codes," *Bell System Technical Journal*, **29**(2): 147, 1950.
3–2. Adams, William, Colonel, USAF Air Materiel Command. Memorandum MCPBXS/WMW/mle to James O. McDonough, transmitting the ruling on patent filings of the Office of The Judge Advocate, Patents and Royalties Division, November 8, 1951. MIT Servomechanisms Laboratory files.
3–3. Aronson, Richard L. "The Forrester Patent," *Control Engineering*, **10**(4): 98–103, 1963.

Chapter 4
4–1. Atwater, Thomas V., Jr, and Robert H. Gregory. "The Economics of Work Performed on a Numerically Controlled Machine Tool," *Journal of Industrial Engineering*, **viii** (6): 337–352, 1957.
4–2. MIT Servomechanisms Laboratory: "Economic Studies of Work Performed on a Numerically Controlled Milling Machine," March 1, 1956.
4–3. Beeby, William D. "Benefits and Economics," *Mechanical Engineering*, **81**(5): 66–67, 1959. Quoted with permission.
4–4. McFerren, Edgar L. (Giddings and Lewis Machine Tool Company). An interview with J. F. Reintjes, Madison, WI, December 9, 1969.
4–5. "Numerically Controlled Machine Tools," MIT Servomechanisms Laboratory notes for an MIT summer session course, August 23 through September 3, 1954.
4–6. Runyon, John H. "Automatic Data Preparation for Numerically Controlled Machine Tools," MIT Servomechanisms Laboratory Engineering Report No. 15 (DIC 6873), January 25, 1956.
4–7. Siegel, Amold. "Information-Processing Routine for Numerical Control," MIT Servomechanisms Laboratory Engineering Report No. 16 (DIC 6873), March 1, 1956.
4–8. Siegel, Amold. "Automatic Programming of Numerically Controlled Machine Tools," *Control Engineering*, **3**(10): 65–70, 1956.
4–9 Ward, John E. Letter to Douglas T. Ross on the APT programming language history, January 26, 1978. MIT Servomechanisms Laboratory files.
4–10. Ross, Douglas T. "Gestalt Programming: A New Concept in Automatic Programming," *Proceedings of the Western Joint Computer Conference*, 1956, American Institute of Electrical Engineers (AIEE). Published as their Special Publication T–85. © 1956 IEEE.
4–11. Ross, Douglas T., and Harry E. Pople, Jr. MIT Servomechanisms Laboratory Interim Engineering Report 6873, IR-1 and IR-2, for the period June 26, 1956 to December 31, 1956.

4–12. Ross, Douglas T. "Origins of the Language for Automatically Programmed Tools," and initial paper presented at SIGPLAN History of Programming Languages Conference, Los Angeles, CA, June 1–3, 1978, pp 3–7. Courtesy of SofTech, Inc., Waltham, MA 02154
4–13. Ross, Douglas T. MIT Servomechanisms Laboratory Memorandum No. 2, "A Proposed Basic Language for the 2D-APT-II," June 14, 1957.

Chapter 5

5–1. Mann, Robert W. "Engineering Specifications for a Man-Computer System for Design," a paper presented at the August 1966 MIT conference on computer-aided design, Cambridge, MA, August 1966.
5–2. Mann, Robert W. and Steven A. Coons. "Computer-Aided Design," *McGraw-Hill Yearbook of Science an Technology*. New York: McGraw-Hill, Inc., 1965.
5–3. Pfeiffer, John. *The Thinking Machine*. Philadelphia and New York: J. B. Lippincott Company, 1962, pp. 130–139.
5–4. Parmelee, Richard P. "A Study of a Stress Analysis for Computer Aided Design," MIT master's thesis, May 22, 1961.
5–5. Parmelee, Richard P. "Three-Dimensional Stress Analysis for Computer Aided Design," MIT doctoral thesis, April 1966.
5–6. Bezier, P. *Numerical Control Mathematics and Applications*. New York: John Wiley & Sons, 1970, pp. 228–235.
5–7. Coons, Steven A. MIT Technical Report MAC-TR-41: "Surfaces for Computer Aided Design of Space Forms," June 1967.
5–8. Hamilton, M. L. and A. D. Weiss. "An Approach to Computer-Aided Preliminary Ship Design," MIT Electronic Systems Laboratory Technical Memorandum ESL-M-228, January 1965.
5–9. Proposal for a research and development program on computer systems made by the Massachusetts Institute of Technology to the U.S. Air Force, Advanced Research Projects Agency (ARPA), January 14, 1963. MIT Servomechanisms Laboratory files.
5–10. Sutherland, Ivan E. "Sketchpad: A Man–Machine Graphical Communication System," a paper presented at the 1963 Spring Joint Computer Conference of the American Federation of Information Processing Societies, Inc. (AFIPS), Vol. 23, pp. 329-345, 1963.
5–11. Johnston, Laird E. "Graphical Communication with a Digital Computer," MIT Master's Thesis, June 1961.
5–12. Gurley, Benjamin M. and Charles E. Woodward. "Light Pen Links Computer to Operator," *Electronics*, **32**(47): November 20, 1959, pp. 85-87.
5–13. Ross, Douglas T. "The AED Approach to Generalized Computer-Aided Design," a paper presented at the August 1967 National Conference of the Association for Computing Machinery (ACM), Washington, DC, pp. 367–385. Also published as MIT Electronic Systems Laboratory Report ESL-R-305, April 1967.
5–14. Katzenelson, J. "Simulation for Nonlinear Circuits by AEDNET," a paper presented at the 3rd Annual SHARE Design Automation Workshop of the Association for Computing Machinery (ACM), New Orleans, LA, May 16–19, 1966, and published in their Proceedings.
5–15. Dertouzos, Michael L. and Charles W. Therrien. "CIRCAL: Online Analysis of Electronics Networks," MIT Electronic Systems Laboratory Report ESL-R-248, October 1965.
5–16. Ross, Douglas T. and John E. Ward. "Investigations in Computer-Aided Design for Numerically Controlled Production." MIT Electronic Systems Laboratory Final Technical Report for the period December 1, 1959 to May 3, 1967.
5–17. Ross, Douglas T., R. H. Stotz, D. E. Thornhill, and C. A. Lang. "The Design and Programming of a Display Interface System Integrating Real-Time and

Time-Shared Computers," a paper presented at the 4th Annual ACM/SHARE Design Automation Workshop, Los Angeles, CA, June 1967, and published in their Proceedings.

5–18. Stotz, Robert H. and Thomas B. Cheek. "A Low-Cost Graphic Display for a Computer Time-Sharing Console," Technical Session Proceedings of the 8th National Symposium on Information Displays of the Society for Information Displays, San Francisco, CA, May 24–26, 1967, pp. 91–100.

Chapter 6

6–1. Webster, William M. Continuous-Path Numerical Control, Paper 58-A-162 presented at the Annual Meeting, New York, NY, November 30–December 5, 1958 of the American Society of Mechanical Engineers. Condensed version published in *Mechanical Engineering*, **81**(5), 68–69, 1959. Quoted with permission

6–2. Goldberg, Alfred, ed. *History of the United States Air Force. 1907–1957.* New York: D Van Nostrand Company, pp. 217-218, 1957.

6–3. Webster, William M. Draft of a United States Air Force position paper on computer-aided design (CAD), June 1966. Courtesy of W. M. Webster. MIT Servomechanisms Laboratory files.

6–4. Pease, William M. Memorandum to Gordon S. Brown on industry response to the numerical control concept, June 6, 1951. MIT, Servomechanisms Laboratory Files.

6–5 "For N/C – A Management Philosophy that Works," *Manufacturing Engineering and Management* pp.57–80, January 1970. The article quotes from an interview with Charles Peterson, Manager, Engine Experimental Fabricating Department, the Ford Motor Company.

6–6. McDonough, James O. and Alfred K. Susskind. Trip report, June 22 through July 2, 1952, submitted to William M. Pease. MIT Servomechanisms Laboratory files.

6–7. Brown, Gordon S. An address on automation in the steel industry given at the 7th International Biannual Convention of the United Steel Workers of America, Atlantic City, NJ, September 1954.

6–8. Brown, Gordon S. and Norbert Wiener. "Automation 1955, A Retrospective," *Annals of the History of Computing*, **6**(4): 372–383, 1984.

6–9. Reintjes, J. F. "Automation–Its Forms and Future," an address at the Third Annual Weekend Conference of the Industrial Relations Section, MIT, and District No.1, United Steel Workers of America, CIO, Cambridge, MA, May 22, 1955. MIT Servomechanisms Laboratory, J. F. Reintjes files.

6–10. Brown, Gordon S. "Automation," *The Technology Reveiw*, **59**(7): 3–8, 1957.

6–11. Reintjes, J. F. "Computer Applications in Automobile Manufacturing," an address at the Second International Skilled Trades Conference on New Technology, United Auto Workers (UAW), Boston, MA, August 17, 1971.

6–12. Reintjes, J. F. Summary of the formal presentation and panel discussion, "Proper Utilization of Technical Personnel in the Nation's Work Force," at a joint Labor Education Conference sponsored by the MIT Employees Union and the American Labor Education Service, May 23–24, 1958. MIT Servomechanisms Laboratory, J. F. Reintjes files.

6–13. Collective bargaining agreement between the Boeing Company and the International Association of Machinists and Aerospace Workers, AFL/CIO, October 4, 1983.

6–14. Morrison, James (Boeing Company). Personal Communication to J. F. Reintjes, May 19, 1987.

6–15. Susskind, Alfred K. An interview with J. F. Reintjes, Cambridge, MA, May, 1969.

6–16. Merner, Kenneth R. (Kearney and Trecker Corporation). Personal communication to J. F. Reintjes, April 9, 1987.

6–17. "Machine Tool Sales Zoom," *Metalworking News*, Machine Tool Show edition, Fairchild Publications, Inc., New York, September 22, 1965, p. 1.
6–18. Transcript of the proceedings of the Founders Convention of the Numerical Control Society, New York, June 14–15, 1962. Courtesy of the Association of Integrated Manufacturing Technology (AIM), Beloit, WI.
6–19. "Numerical Control Society is Formed," *American Machinist*, **106**(13): 66, 1962.

Chapter 7

7–1. Wood, Kenneth G. (Boeing Company). "Problems Associated with the Introduction of Numerical Control into Aircraft Plants – from an Industrial Viewpoint," a lecture presented at MIT on April 2, 1957 as part of a Servomechanisms Laboratory sponsored special course on Programming for Numerically Controlled Machine Tools. MIT Servomechanisms Laboratory files.
7–2. Petrov, V. "Machine Tools with Program Controls are at a Standstill," *Leningradskaya Pravda*, July 28, 1962. Translation by US Department of Commerce and reprinted in its Bulletin No. 36: USSR Industrial Development, "Soviet Metalworking Equipment," November 6, 1962.
7–3. Merchant, M. Eugene (Cincinnati Milacron, Inc.). "Future Trends in Manufacturing–Toward the Year 2000," a keynote paper presented at a conference on Technology and Innovation for Manufacturing, Washington, DC, May 4–5, 1979, sponsored by the Committee on Science and Technology of the US House of Representatives, 96th Congress (Second Session), and published in their Proceedings. US Government Printing Office, February 1980.
7–4. Bruce S. Olds Associates, Inc. "Return on Investment in Basic Research – Exploring A Methodology," a report to the Office of Naval Research, Department of the Navy, under Contract N00014-79-C-0192, November 1981.

Afterword

A–1. Wagoner, Harless D. *The US Machine Tool Industry*, Cambridge, MA: The MIT Press, p. 61, 1966.
A–2. Ibid, pp. 362, 363.
A–3. Melman, Seymour. "How the Yankees Lost Their Know-How," *MIT Technology Review*, pp. 56–64, October 1983.
A–4. Noble, David F. *Forces of Production*. New York: Alfred A. Knopf, pp. 104–143, 1984.
A–5. Reference A–3, p. 58.
A–6. Letter to Ralph E. Cross from J. F. Reintjes, February 15, 1955.
A–7. Letter to J. F. Reintjes from Ralph E. Cross, March 14, 1955.
A–8. Dertouzos, Michael, Richard K. Lester, Robert M. Solow, *Made in America*. Cambridge, MA: The MIT Press, 1989.

BIBLIOGRAPHY

The literature abounds in documents on all aspects of numerical control. The following list represents a partial collection beyond the citations in the *References* section of this book. It has been put together with emphasis on the early years of the technology and with some attention to documents which are themselves bibliographies or which contain a substantial number of bibliographic references. Items within each category are ordered chronologically.

Books and Conference Proceedings

Susskind, Alfred K. *Notes on Analog–Digital Conversion Techniques*. Cambridge, MA: MIT Technology Press, 1957.

Proceedings of the Symposium on Numerical Control Systems for Machine Tools, sponsored by the Electronic Industries Association (EIA), Los Angeles, CA, September 17–18, 1957. Published by New York Engineering Publishers, a Division of the AC Book Company.

American Society of Tool and Manufacturing Engineers (ASTME). *Numerical Control in Manufacturing*. New York: McGraw-Hill, Inc., 1963.

Wagoner, Harless D. *The US Machine Tool Industry from 1900 to 1950*. Cambridge, MA: The MIT Press, 1968.

Horton, H. E. *Numerical Control of Machine Tools*. New York: Brandon/Systems Press, Inc., 1969. [A collection of numerical control words and terms with their equivalent expressions in English, French, German, and Spanish.]

Howe, Raymond E., ed. *Introduction to N/C in Manufacturing*. Dearborn, MI: ASTME, 1969.

Leslie, W. H. P., ed. *Numerical Control Users Handbook*. Maidenhead, England: McGraw-Hill, Inc., 1970.

Proceedings of the Numerical Control Society's Annual Meetings, 1970–1983, Beloit, WI: Association of Integrated Manufacturing (AIM).

Olesten, Nils O. *Numerical Control*. New York: John Wiley & Sons, 1970.

Patton, William J. *Numerical Control: Practice and Application*. Reston, VA: Reston Publishing Company, 1972.

Harrington, Joseph, Jr. *Computer Integrated Manufacturing*. New York: Industrial Press, 1973.

McPherson, D., ed. *Advances in Computer-Aided Manufacture*. New York: North-Holland, 1976. [Proceedings of the 3rd International IFIP/IFAC Conference on Programming Languages for Machine Tools, Stirling, Scotland, June 15–18, 1976]

Morse, Philip M. *In at the Beginning: A Physicist's Life*. Cambridge, MA: The MIT Press, 1977.

Pressman, Roger S. and John E. Williams. *Numerical Control and Computer-Aided Manufacturing*. New York: John Wiley & Sons, 1977.

Koren, Yoram and Joseph Ben-Uri. *Numerical Control of Machine Tools*. Delhi, India: Khana Publishers, 1978.

Childs, James J. *Principles of Numerical Control*, 3rd. ed. New York: Industrial Press, Inc., 1982.
Gibbs, David. *An Introduction to CNC Machining*. Eastbourne, East Sussex, England: Castle, Ltd, 1984.
Noble, David F. *Forces of Production: A Social History of Automation*. New York: Alfred A.Knopf, Inc., 1984.
Crestin, J. P., and J. F. McWaters (eds). *Software for Discrete Manufacturing*. Amsterdam: North-Holland, 1986; New York: Elsevier Science Publishing Company, sole distributors for USA and Canada. [Proceeding of the 6th International IFIP/IFAC Conference on Software for Discrete Manufacturing, Paris, 1985].

Reports

MIT Servomechanisms Laboratory. "APT", vols. I-II; IV-VII. [A complete documentation of the Automatically Programmed Tool system as of 1959, authored by various staff members of Servo Lab's Computer Applications Group.]
"Report of the Long-Range Study Group," MIT Servomechanisms Laboratory, J. F. Reintjes files, April, 1961.
US Department of Labor, "Occupations in Numerically Controlled Metal-Cutting Machinery," Superintendent of Documents, US Government Printing Office, Washington, DC 20402, 1968.
Ward, John E. "Computer-Aided Design for Numerically Controlled Production," MIT Electronic Systems Laboratory, Technical Report AFML-TR-70-78, June 1970.
McKenzie, John A. "TX-O History," MIT Research Laboratory of Electronics, October 1, 1974.
Ross, Douglas T. and John E. Ward. "Investigations in Computer-Aided Design for Numerically Controlled Production," MIT Electronic Systems Laboratory, Final Technical Report, December 1, 1959 to May 3, 1967. [Includes a bibliography of 274 items published by project members from 1959 to 1967.]

Articles, Booklets and Pamphlets

Pease, William M. "An Automatic Machine Tool," *Scientific American*, **187**(3): 101–115, 1952.
Jacoby, C. J., 111. Analysis of Developments in Automation," *Mechanical Engineering*, **74**(10): 810–811, 1952.
McDonough, James O. and Alfred K. Susskind. "A Numerically Controlled Milling Machine," Proceedings of the Joint Eastern Computer Conference of the American Institute of Electrical Engineers (AIEE), the Institute of Radio Engineers (IRE) and the Association of Computing Machinery (ACM), AIEE Publication S-53, 1953, pp. 133–136.
McDonough, James O. "Punched Tape Guides Milling Machine Cutters," *Electronics*, **26**(4): 133–136, 1953.
Mergler, Harry W. "A Digital–Analog Machine Tool Control System," Proceedings of the Western Computer Conference, Los Angeles, CA, February 11–12, 1954, AIEE Publication S-59, 1954, pp. 46–59.
Cunningham, R. W. "Controlling Machine Tools Automatically," *Mechanical Engineering*, **76**(6): 487–490, 1954.
Stocker, W. M., Jr. and C. D. Emerson. "Numerical Control," *American Machinist*, **98**(22): 133–156, 1954.
"New Tools... Automated, Make '55," *American Machinist*, **98**(23): 157–164, 1954.
Kelling, Leroy U. C. "Numerical Control of Machine Tools," *Institute of Radio Engineers, Transactions on Industrial Electronics*, PGIE-2: 3–8, March 1955.
Brown, J. H. "Measurement Motion to 0.0001 in. without Friction or Wear," *Control Engineering*, **2**(4): 50–52, 1955.
Williamson, D. T. N. "Computer-Controlled Machine Tools," *Aircraft Production*, **17**(7): 262–278, 1955.

Eshelman, Ralph H. "Numerical Machine Control," *The Tool Engineer*, **35**(4): 117–126, 1955.

Williamson, D. T. N. "Applying Electronic Computer Controls," *Machine Design*, **27**(10): 238–242, 1955.

"Computer-Prepared Numerical Data for Machine Control," *Electrical Manufacturing*, **58**(1): 102–109, 1956.

Findlay, David A. "Electronic Controls for Machine Tools," *Electronics*, **29**(1): 122–129, 1956.

Farmer, Peter J. "Analogue Control, Application of E.M.I. Control to a Standard Vertical Milling Machine," *Aircraft Production*, **18**(4): 126–134, 1956.

Farmer, Peter J. "Digital Control" *Aircraft Production*, **18**(7): 256–267, 1956.

Jaeger, J. J. "Numerical Control for Jig Boring Holds Production Promise," *The Tool Engineer*, **37**(1): 102–107, 1956.

Johnson, E. C. "A Numerically Controlled Cam-Milling Machine," *Institute of Radio Engineers, Transactions on Industrial Electronics*, PGIE-3: 80–86, March 1956.

Johnson, E. C. "Numerically Controlled Milling," *Automation*, **3**(6): 60–64, 1956.

Mergler, Harry W. "Machining from Recorded Information," *Control Engineering*, **3**(9): 110–111, 1956.

Susskind, Alfred K. "Digital Information Processing for Machine-Tool Control," Institute of Radio Engineers Convention Record, **5**(4): 1957, pp. 145–149.

"APT – The Language that Commands Machine Tools," *American Machinist*, **103**(5): 106–107, 1959.

Mann, Robert W. "Computer-Aided Mechanism Design," Paper No. 64-Mech-36, ASME Mechanisms Conference, Lafayette Ind., October 19–21, 1964.

"Numerical Controls," US Department of Commerce, National Bureau of Standards, (Clearinghouse for Federal Scientific and Technical Information, Springfield, VA 22151), May 1965. [An annotated bibliography on numerical control from 1956 to 1964. Several references are themselves bibliographies in which are cited some sixteen hundred references on data processing, programming and metal-working.]

Coons, Steven Anson. "The Uses of Computers in Technology," *Scientific American*, **196**(9): 177–188, 1966.

Meissner, Charles W., Jr., "An Annotated Bibliography of Computer Aided Analysis and Design," Scientific and Technical Information Division, Office of Technology Utilization, National Aeronautics and Space Administration, Washington, DC, 1968.

Reintjes, J. F. "Crucial Decisions during the Evolution of Numerical," a paper presented at the Annual Meeting of the Society for the History of Technology, Washington, DC, October 20–22, 1977. MIT Electronic Systems Laboratory, J. F. Reintjes files.

Manufacturing Engineering, Golden Anniversary Issue, 1932–1982, Society of Manufacturing Engineers, **8**(1): 1982.

Ward, John E. "The Technical Development of Numerical Control," a paper presented at the Annual Meeting of the Society for the History of Technology, Washington, DC, October 20–22, 1977. MIT Electronic Systems Laboratory, J. F. Reintjes files.

Films

The following films are at the MIT Museum, Cambridge, MA.

A Numerically Controlled Machine Tool. A 20-minute, color, sound, 16-mm film (1963). A lucid exposition of the basic principles of numerical control of machine tools. The film describes the first numerically controlled machine tool, developed in 1952 at the MIT Servomechanisms Laboratory.

APT. A 30-minute, black-white, sound 16-mm film (1959). A film of an educational television program describing the Automatically Programmed Tool system.

Thinking Machine. A 60-minute film (1961), which appeared as part of the CBS "Tomorrow" series of television programs.

Kludge. A 7-minute, black-white, silent, 16-mm film (1966). Shown are the Project MAC computer, the ESL Display Console (the "Kludge"), two- and three-dimensional drawings made with the aid of the light pen, and three-dimensional real-time rotation of figures and surfaces.

CIRCAL-O. A 7 minute, black-white, silent, 16-mm film (1966). This film shows how an on-line digital computer utility and an electronic circuit designer "cooperate" through a program that analyzes electronic networks.

Sketchpad. A 7-minute, black-white, sound, 16-mm film (1962), illustrating a two-dimensional graphical language with constraints.

APPENDIX A

Memorandum of Agreement between the Massachusetts Institute of Technology and the Parsons Corporation

MASSACHUSETTS INSTITUTE OF TECHNOLOGY
DIVISION OF INDUSTRIAL COÖPERATION
CAMBRIDGE 39, MASS.

N. McL. SAGE
DIRECTOR

F. L. FOSTER
ASSISTANT TO THE DIRECTOR

DIC 6694

MEMORANDUM OF AGREEMENT between the
MASSACHUSETTS INSTITUTE OF TECHNOLOGY, hereinafter called "the Institute," and PARSONS CORPORATION, Traverse City, hereinafter called "the Sponsor."

The Institute will undertake a research program for the Sponsor as follows:

1. THE FIELD OF RESEARCH comprises two phases as follows:

Phase I - An engineering study directed toward the development of a system applicable to machine tools for controlling the position of shafts in accordance with the output of a computing machine.

Phase II - The supply of an apparatus to operate a demonstration machine tool to hold a performance accuracy so that the surface finish has no roughness in excess of .001 inch, and is correctly located within .003 inch when operating at a commercial table-feed rate.

2. SPONSOR'S COMMITMENTS. The Sponsor will pay the cost of the research, exclusive of patent costs, upon monthly billings by the Institute. This cost includes:

(a) salaries and wages of staff properly chargeable to the project at the following hourly rates:

Executive staff	$8.50
Principal engineers	6.50
Engineers	4.50
Mechanics, laboratory technicians, drafting personnel	4.00
Nontechnical personnel (direct stenographic, and guards when required for security)	2.75

These rates shall be subject to review on January 1, 1950. They are computed to return an amount approximately equal to direct salaries and wages, which enables our commercial work to make a contribution to the Institute to help it maintain its facilities and research position.

(b) the cost of all materials, services, and equipment contracted for by the project.

3. INSTITUTE'S COMMITMENTS. The research will be under the supervision of Dr. Gordon S. Brown. The cost of the research under item (2) above will be established in a mutually agreed upon budget which the Institute will not exceed without specific written authorization or modification of the budget to take into account new conditions.

4. PATENTS. The Institute will require all employees engaged in the research to execute agreements by which they undertake to communicate inventions to

the Sponsor, to execute patent applications and to assign to the Sponsor such applications and any patents that may issue thereon.

Inventions made by persons employed on the project in carrying out the work contemplated by this agreement shall be handled as follows:

(a) The Sponsor will have the right to file and prosecute at its expense patent applications in the name of the inventor but with assignments to the Sponsor. The Sponsor will keep Institute counsel informed of all applications filed, and actions and amendments therein.

(b) In the field of the research, the Sponsor shall have full rights, except that after ten (10) years from the date of filing of any application the Institute will have the free right to grant licenses to others thereunder.

(c) In the event inventions are made having uses outside the field of research, as to such uses the Institute will have full rights including the sole right to license others, except that the Sponsor will have the free right to employ said inventions in its own business or in the business of any subsidiary or parent companies of the Sponsor.

(d) Should the Sponsor elect not to file on any invention or decide to abandon the prosecution of a patent application, the Institute will have the right to file or take over prosecution at its own expense, in which case the Sponsor will receive only a royalty-free, non-exclusive license thereunder.

Any assignment of rights by either party shall be subject to the foregoing.

5. PERIOD OF THE PROJECT. This project will continue for one year from July 1, 1949 and will be subject to renewal only by mutual agreement.

6. TERMINATION. This work may be terminated by either party on thirty days written notice. In the case of termination by the Sponsor, the Institute will be reinbursed for all commitments under sections 2(a) and 2(b).

7. USE OF THE NAME OF THE INSTITUTE. The Sponsor agrees not to use the name of the Institute or of any member of its staff in sales promotion work or advertising, or in any other form of publicity without the written permission of the Director of the Division of Industrial Cooperation.

8. PUBLICATION Subject to restrictions places on publication by security the Institute will be free to publish papers after giving a copy of the publication to the Sponsor.

Place and date ____________ MASSACHUSETTS INSTITUTE OF TECHNOLOGY

______________________ By ______________________

Witnesses: ______________________

____

______________________ By ______________________

APPENDIX B

The 1952 Demonstration of the MIT Numerically Controlled Milling Machine

Servomechanisms Laboratory

of the

Massachusetts Institute of Technology

under a program sponsored by

Air Materiel Command

U. S. Air Force

will demonstrate a

Numerically Controlled Milling Machine

at M. I. T.

September 15, 16, 17, 1952

TO ACQUAINT MEMBERS *of the aircraft, machine-tool, and other industries with an important new development in automatic control of machine tools, a series of demonstrations and discussions of the M.I.T.*

Numerically Controlled Milling Machine

will be held at the Massachusetts Institute of Technology on September 15, 16, and 17, 1952.

UNDER SPONSORSHIP of the Air Materiel Command, U. S. Air Force, the Servomechanisms Laboratory of the Massachusetts Institute of Technology completed in March of 1952 the construction of a milling machine capable of manufacturing machined parts automatically by obeying a series of numerical instructions introduced into the machine on punched paper tape. Since its completion, the machine has been subjected to a series of engineering tests and has performed sample work in a variety of areas. It is now possible to make a preliminary assessment of the economic potential of this new form of machine-tool control. It is therefore appropriate to acquaint industry with this development and to initiate an interchange of information in order to guide the work now in progress at M.I.T. along lines which will result in maximum usefulness to industry.

A series of one-day demonstrations will be given at M.I.T. to which industrial representatives are invited. While all three demonstrations are essentially alike, it is desired that guests having similar interests attend on the same date.

On Monday, September 15, the demonstration and discussion will be directed toward the phases of this development of particular interest to machine-tool builders. Members of the National Machine Tool Builders' Association are especially invited to this session.

On Tuesday, September 16, the program will be less specifically directed toward machine-tool builders or aircraft manufacturers. Persons interested in other applications of automatic controls are especially invited to this session.

On Wednesday, September 17, the application of numerical control to the production of aircraft parts will be discussed. Members of the Aircraft Industries' Association are especially invited to this session.

REGISTRATION

SPACE LIMITATIONS AT M.I.T. require that company representation be limited to one or two persons and that all persons desiring to attend one of the sessions return the enclosed registration blank prior to September 8 to:

Servomechanisms Laboratory
Building 32
Massachusetts Institute of Technology
Cambridge 39, Massachusetts
Attention: Mr. J. O. McDonough

PROGRAM

Monday, September 15, 1952

"Numerical Control and the Machine-Tool Builder"
Huntington Hall
Massachusetts Institute of Technology

9:15 A.M.	Registration of Guests in Lobby, 77 Massachusetts Avenue, Cambridge, Massachusetts
10:00 A.M.	Welcoming Remarks
10:15 - 11:00 A.M.	Orientation
11:00 A.M. - 3:30 P.M.	Meetings on the Following Subjects: Machine Demonstration; Tape Preparation; Sample Workpieces
3:30 - 4:00 P.M.	Modern Information Processing
4:00 - 5:30 P.M.	General Discussion on Numerical Control
8:00 P.M.	Informal Discussion

Tuesday, September 16, 1952

"Industrial Applications of Numerical Control"
Huntington Hall
Massachusetts Institute of Technology

9:15 A.M.	Registration of Guests in Lobby, 77 Massachusetts Avenue, Cambridge, Massachusetts
10:00 A.M.	Welcoming Remarks
10:15 - 11:00 A.M.	Orientation
11:00 A.M. - 3:30 P.M.	Meetings on the Following Subjects: Machine Demonstration; Tape Preparation; Sample Workpieces
3:30 - 4:00 P.M.	Modern Information Processing
4:00 - 5:30 P.M.	General Discussion on Numerical Control
8:00 P.M.	Informal Discussion

Wednesday, September 17, 1952

"Numerical Control and the Aircraft Industry"
Huntington Hall
Massachusetts Institute of Technology

9:15 A.M.	Registration of Guests in Lobby, 77 Massachusetts Avenue, Cambridge, Massachusetts
10:00 A.M.	Welcoming Remarks
10:15 - 11:00 A.M.	Orientation
11:00 A.M. - 3:30 P.M.	Meetings on the Following Subjects: Machine Demonstration; Tape Preparation; Sample Workpieces
3:30 - 4:00 P.M.	Modern Information Processing
4:00 - 5:30 P.M.	General Discussion on Numerical Control
8:00 P.M.	Informal Discussion

Organizations and Number of Representatives at the Numerically Controlled Machine Tool Demonstration September 15–17, 1952

Abrasive Machine Tool Company (1)
Air Force:
—Air Materiel Command:
Plant Representatives (4)
Procurement (10)
Wright Patterson Air Force Base (5)
—Cambridge Research Center (2)
Aircraft Industries Association (1)
Aluminum Company of America (2)
American Steel and Wire Company (1)
American Steel Foundries (1)
Arter Grinding Machine Company (2)
Audio-Video Corporation (1)
Avey Machine Drilling Company (2)
Barber–Coleman Company (1)
Bardons and Oliver, Inc. (1)
Bell Aircraft Corporation (1)
Bell Telephone Laboratories, Inc. (1)
Bendix Aviation Corporation (2)
Blanchard Machine Company (2)
Boeing Airplane Company (2)
Boye and Emmes Machine Tool Company (1)
Bridgeport Machines, Inc. (2)
Brown and Sharp (2)
Bryant Chucking Grinder Company (1)
Business Week Magazine (1)
Capehart–Farnsworth Company (2)
Chance–Vought Aircraft (1)
Cincinnati Gilbert Machine Tool Company (2)
Cincinnati Milling Machine Company (4)
Clark Equipment Company (2)
Consolidated Vultee Aircraft Corporation (3)
Continental Motors Corporation (1)
Curtiss–Wright Corporation (3)
Denison Engineering Company (2)
Dennison Manufacturing Company (1)
Department of Defense Munitions Board (2)
Douglas Aircraft Company (1)
Draper Corporation (2)
DuPont (1)
Emhart Manufacturing Company (1)
Excello Corporation (1)
Farrel–Birmingham Company, Inc. (3)
Fellows Gear Shaper Company (3)
Fenwal, Inc. (1)
Fitchburg Engineering Corporation (1)
Fosdick Machine Tool Company (1)
Fox Engineering Company (1)
General Alloys Company (1)
General Electric Company (6)
General Machinery Corporation (2)
General Motors Corporation (1)
George Gorton Machine Company (2)
Giddings and Lewis Machine Tool Company (2)
Gisholt Machine Company (1)
Grumman Aircraft Engineering Corporation (2)
Hanson and Whitney Company (1)
Harvard College (1)
Heald Machine Company (1)
Hendey Machine Company (4)
Himoff Machine Company (2)
Hollingsworth and Whitney (1)
Hughes Aircraft Corporation (1)
Ingersoll Milling Machine Company (1)
International Business Machines Corporation (3)
Jones and Lamson Machine Company (3)
Kaman Aircraft Corporation (1)
Kaiser Manufacturing Corporation (3)
Kendall Mills (2)
W. B. Knight Machinery Company (1)
Landis Tool Company (2)
Leland–Gifford Company (1)
Liberty Products Corporation (2)
Arthur D. Little, Inc. (1)
Lobdell United Company (1)
Lockheed Aircraft Corporation (2)
Lodge and Shipley Company (1)
W. L. Marson Corporation (2)
Glenn L. Martin (3)
Miniature Precision Bearings, Inc. (2)
Minneapolis-Honeywell Regulator Company (2)
Monarch Machine Tool Company (1)
Robert E. Morris Company (2)
Morton Manufacturing Company (1)
National Automatic Tool Company, Inc. (2)
New Hampshire Ball Bearrings, Inc. (2)
North American Aviation, Inc. (2)
Norton Company (3)
Notch and Merryweather Machinery Company (1)

Oerliken Tool and Arms Corporation of America (2)
Olin Industries, Inc. (1)
Parsons Corporation (2)
Philco Corporation (1)
Pittsburgh Plate Glass Company (1)
Robert R. Podgers (1)
Potter and Johnston Company (2)
Pratt and Whitney (4)
Raytheon Manufacturing Company (5)
Reed–Prentice Corporation (2)
Republic Aviation Corporation (1)
Rohr Aircraft Corporation (1)
Saco Lowell Shops (2)
Seneca Falls Machine Company (1)
Sheffield Corporation (2)
Sikorsky Aircraft Division (1)
A. O. Smith Company (3)
Sperry Gyroscope Company (2)
Sprague Electric Company (1)
Standard Oil Company (1)
Sunstrand Machine Tool Company (1)
Taft-Pierce Manufacturing Company (2)
Thompson Grinder Company (2)
Thompson Products, Inc. (2)
Tower Iron Works (2)
Ultrasonic Corporation (2)
United Aircraft Corporation (2)
United Shoe Machinery Corporation (1)
United States Steel Company (1)
Van Norman Company (1)
Verson Allsteel Press Company (1)
Warner and Swasey Company (2)
Waterbury Tool Company (1)
Western Electric Manufacturing Company (1)
Westinghouse Electric Corporation (7)
Wiesner–Rapp Company, Inc. (2)
Wyman–Gordon Company (2)

APPENDIX C

Rosters of Attendees at the 1954, 1955 and 1957
Two-Week Intensive Courses on Numerical Control

Monday, August 23 through Friday, September 3, 1954

American Machine & Foundry
 Jules E. Kadish
American Optical Company
 Harry N. Lanza
Ampex Corporation
 James B. Norcross
Audio & Video Products Company, Inc.
 James U. Lemke
Bell Telephone Laboratories
 Frank W. Kinsman
Cincinnati Milling Machine Company
 Joseph P. Ivanska
Citroen Cars Manufactory
 Georges de la Boixiere
Convair
 Melburn C. Copold
Cross Company
 Alfred P. Langlois
Cumberland Steel Company
 W. Wallace McKaig
Curtis-Wright Corporation
 George W. Walker
Eastman Kodak Company
 John H. Juengst
Elgin National Watch Company
 Richard G. Schimpf
Ex-cell-O Corporation
 Francis Vargo
Farrand Optical Company, Inc.
 Vincent F. Foster
 James Winget
Federal Products Corporation
 Herbert A. Johnson
Fellows Gear Shaper Company
 David E. Armstrong
 Stuart G. Leonard
General Electric Company
 Leroy E. Foley
 Philip D. Kaplan
 Lester E. McCabe
 Richard J. Rieke
General Motors Corporation
 Wayne E. Shannon
General Riveters, Inc.
 Charles R. Alford
Giddings & Lewis Machine Tool Company
 Walter L. McCann
Gisholt Machine Company
 Edward C. Helmke
I.B.M.
 Donald H. Apgar
 Rex C. Bradford
 Kurt Eisemann
 Mark Morgan
Jones & Lamson Machine Company
 Walter Shultz
Kaiser Metal Products, Inc.
 Anthony DiGuiseppe
Kearney & Trecker Corporation
 John R. Joerger
Glenn L. Martin
 Joseph F. Hayward
 William L. Lambdin
Minneapolis-Honeywell Regulator Co.
 Theodore B. Gurniak
Morey Machinery Company, Inc.
 Joel Feiner
New York Air Brake Company
 Max Luming
North American Aviation, Inc.
 Oliver A. Foss
Norton Company
 George E. Comstock, III
Oerlikon Tool & Arms Corporation
 Stanley Breen
Radio Shack
 Edmund J. Keane
Raytheon Manufacturing Company
 Harvey A. Miller
Sperry Gyroscope Company
 Richard J. Monaco
Stromberg-Carlson Company
 Howard S. Gleason
Sylvania Electric Products, Inc.
 JosephC. Siedlecki
Thompson Products, Inc.
 Charles O. Parratt
Ultrasonic Corporation
 Arthur W. Dickson
United Shoe Machinery Corporation
 Robert R. Perron
Watertown Arsenal
 Ellsworth H. Swift
Warner & Swasey
 Ralph H. Schuman
Western Electric Company
 Elmer N. Lenk
Westinghouse Electric Company
 W. Marshall Brittain
 James E. Frederick
 Alexander R. Nelson
 Robert I. Van Nice
Other
 Peter G. Buehning
 William L. Merrit

SPECIAL SUMMER PROGRAM 6.601 NUMERICAL CONTROL OF MACHINE TOOLS

Monday, August 22 through Friday, September 2, 1955

Bendix Aviation Corporation
Earl R. Beck, Jr.
E. W. Bliss Company
Richard L. Curtner
Boeing Airplane Company
Howard B. Buchanan
Edward F. Carlberg, Jr.
William H. McKnight
William W. Wilder
John A. Wilkening
W. William Woods
Burroughs Research Center
Anthony L. Consalvi
George A. Harris
Paul W. Winkelman
The Carborundum Company
Richard D. Rutt
Clevite Research Center
David C. Caulkins
Continental Can Company, Inc.
Joseph J. Sennello
Crosley Research Laboratory
Aaron L. Keller
Data Processing Associates, Ltd.
George S. Glinski
Frankford Arsenal
Francis X. McGarrigle
General Motors Corporation
William A. Fletcher
James F. Salatin
Giddings and Lewis Machine Tool Co.
Robert G. Chamberlain
Guy E. Lingenfelter
The Heald Machine Company
Charles F. Carter
I.B.M.
Richard Bartsch
Harold W. Brainard, Jr.
Lawson V. Griffin
Iowa State College
L. Sigfred Linderoth, Jr.
Kearney and Trecker Corporation
Theodore A. Wetzel
Lockheed Aircraft Corporation
R. Lowell Hand
Glenn L. Martin Company
Stanley A. Duran
Metcut Research Associates
Norman Zlatin
Minneapolis-Honeywell Regulator Co.
Royal R. Hawkins
M.I.T.
B.O. Lindahl
Akira Watanabe
Morey Machinery Company, Inc.
Joel Feiner
National Automatic Tool Company
Virgil E. Tice
North American Aviation, Inc.
Lionel S. Peck
David H. Ross
Oerlikon Tool and Arms Corporation
Robert A. Wade
The Plessey Company, Ltd.
Harold J. Finden
Radio Corporation of America
Francis J. Yannotti
Reliance Electric and Engineering Company
Robert W. Pfaff
Remington Rand
Alan E. Stillman
Rockford Machine Tool Company
Ralph S. Litterst
Sheffield Corporation
Charles H. Laemmel
Sperry Gyroscope Company
Peter J. Isaacs
Standard Telecommunications Labs., Ltd.
John K. Webb
Stromberg-Carlson Company
John Fondrk
Sunstrand Machine Tool Company
Carl F. Erikson
Sylvania Electric Products, Inc.
Edward F. Gallagher
United States Air Force
Captain William H. Lynch
Lincoln E. Walker
William M. Webster
U.S. Naval Gun Factory
Moe I. Schneebaum
Paul Tarver
University of Pennsylvania
Julius Tou
Westinghouse Electric Corporation
Edgar L. Juell, Jr.
Jack D. Meess
James E. Stell

SPECIAL SUMMER PROGRAM 6.609 PROGRAMMING FOR NUMERICALLY CONTROLLED MACHINE TOOLS

Monday, March 2 through Wednesday, April 3, 1957

Avco, Lycoming Division
- Ernest C. Goudreau

Battelle Memorial Institute
- Roger C. Van Buskirk

Boeing Airplane Company
- Lloyd H. Bradshaw
- Charles E. Butzberger
- Samuel T. Byars
- Edward F. Carlberg
- Orel R. Evans
- Bert E. Forsman
- Edward E. Graham
- Lyle R. Mhyre
- Robert W. Miller
- Richard E. Stalcup
- Arthur D. Wheless

Canadair, Ltd.
- Dennis H. Brown

Cleveland Pneumatic Tool Company
- Fred W. Olsen

Chance Vought Aircraft, Inc.
- Albert H. Blacker

Convair
- Roland I. Hagar
- Hal R. Montgomery
- James B. Rankin
- Charles B. Young

Curtis-Wright Corporation
- Stanley L. Samuels

Douglas Aircraft Company
- Walter J. Kinney
- Harold F. Robinson

Ford Motor Company
- L. E. Ording

Goodyear Aircraft Corporation
- Fernand Belanger
- George R. Boucher
- Arvel D. Sharp

General Electric Company
- Michael J. Lacy

Lockheed Aircraft Corporation
- Joseph Albert
- Alan D. Baumbach
- Victor Benatar
- Elliott A. Green
- Henry M. Lakin
- Sheldon S. Messick
- Henry P. Philpot
- James B. Tyson

Glenn L. Martin Company
- Elwood B. Dohner
- Frederick W. Mowry
- Howard M. Peckham
- Berton A. Richardson
- Clarence W. Trowbridge
- Vincent M. Werbach

McDonnell Aircraft Corporation
- James A. Horn

Marquardt Aircraft Company
- Kenneth C. Sanderman
- George A. Steven

North American Aviation, Inc.
- David H. Ross

Northrop Aircraft, Inc.
- Arthur L. Eskelin
- Wylye P. Robertson, Jr.
- Thruston M. Rothrock

Republic Aviation Corporation
- James J. Childs

Rohr Aircraft Corporation
- Carl F. Quick
- Kenneth C. Roye

Sperry Gyroscope Company
- Murray Lehrer

Thompson Products, Inc.
- Charles O. Parratt
- Robert Oliver Young

United Aircraft Corporation
- Charles E. Robinson

Westinghouse Electric Corp.
- Henry A. Leone

APPENDIX D

An Outline of Subject Matter for Industrial Training Programs in Numerical Control of Machine Tools

AN OUTLINE of SUBJECT MATTER for INDUSTRIAL TRAINING PROGRAMS in NUMERICAL CONTROL of MACHINE TOOLS

Origin of the Outline

During the period March 25 through April 3, 1957, representatives of companies interested in Numerical Control participated in a special course entitled PROGRAMMING FOR NUMERICALLY CONTROLLED MACHINE TOOLS at the Massachusetts Institute of Technology, Cambridge, Massachusetts. As part of the course, the participants undertook the formulation of an outline which might serve as a guide to industrial organizations planning to set up company training programs in Numerical Control, with emphasis on familiarization, programming, and data processing. The outline is presented in this leaflet.

How to Use the Outline

The working groups, in preparing the outline, recognized that because departmental structures, training methods, and number and backgrounds of trainees may differ greatly among companies, no single course or set of courses could be synthesized to meet exactly the needs of all. Accordingly, emphasis was placed upon outlining a body of subject matter, which, in the judgement of the authors, embraces most aspects of programming for Numerical Control. It is suggested that companies using the outline should be prepared to rearrange, delete, or expand upon the material in accordance with their specific needs. The development of training courses in maintenance and operation of numerically controlled machine-tool systems that are commercially available is left to the individual companies.

Many of the subject headings in the outline are rather broad in scope. The authors intended them to be this way. Because Numerical Control is a relatively new field, frequent advancements in its techniques may be expected. The task of interpreting each topic in terms of current knowledge and experience is left to the instructor.

PRINCIPLES OF NUMERICAL CONTROL

INTRODUCTION

DEFINITION OF NUMERICAL CONTROL

(for example, it is the automatic controlling of processes or machines in accordance with instructions expressed as numbers)

STATUS

History

EARLY

examples such as looms, player piano

RECENT

the M.I.T. developments

Present

VARIOUS COMMERCIAL POSITIONING SYSTEMS

VARIOUS CONTOURING SYSTEMS

examples such as Bendix, Concord Control Inc. (director only), Contraves, E.C.S., E.M.I., Farrand, Ferranti, General Electric, North American, Teller

Future

APPLICATIONS

HARDWARE IMPROVEMENTS

RELIABILITY CONSIDERATIONS

ECONOMIC CONSIDERATIONS

TOOLING

MATERIAL (SCRAPPAGE)

PRODUCT QUALITY

MACHINE TIME

MAKE-READY TIME

FLOOR SPACE

CAPITAL COST

Machine and controls

Data processing equipment

Spare parts and testing equipment

TRAINING
MAINTENANCE
DATA PROCESSING
EFFECTS ON SKILL REQUIREMENTS AND ORGANIZATIONS

COMPONENTS

DATA-PROCESSING EQUIPMENT

General purpose
Special purpose

MACHINE-TOOL HARDWARE

DESCRIPTION OF TYPES OF SYSTEMS

POSITIONING (POINT TO POINT)

Machine tool
Auxiliary equipment
Applications

CONTOURING (CONTINUOUS PATH)

Machine tool
Auxiliary equipment
Applications

AFFECTED FUNCTIONAL AREAS

PRODUCT DESIGN
DRAFTING PRACTICES
METHODS
DATA PROCESSING
TOOLS
MACHINE OPERATIONS
INSPECTION
MAINTENANCE
LABOR RELATIONS

METHODS AND PROGRAMMING

The range of the methods and programming phase of the outline extends from the receipt of the blueprint by the manufacturing planning department to the final preparation of the manuscript. It is felt that details of preplanning instruction, as well as the actual "follow through" on fabrication of the part, should be left to the individual company. The reason for this is that a wide variance of organization, structure, and policy exists among companies.

MACHINE CAPABILITIES AND LIMITATIONS

- AXES OF MOTION CONTROLLED (AND DISTANCES)
- FEEDS AND SPEEDS
- HORSE-POWER SPECIFICATIONS
- HEADS
- FUNCTIONS CONTROLLED – COOLANT, CLAMP, *and so forth*
- SIZE (GENERAL DIMENSIONS)
- DYNAMICS OF THE MACHINE – ACCELERATION, DECELERATION
- WORKING CLEARANCES
- WORKING AREA (TABLE SIZE, *and so forth*)
- ACCURACIES
- TRAVEL LIMITS
- RAPID TRAVERSE DATA
- SPINDLE SIZE AND ROTATION
- MANUAL CONTROLS – CONVENTIONAL MOTIONS
- CUTTING TIME FOR SPECIFIC AMOUNT OF TAPE
- SAFETY FEATURES
- SPECIAL MACHINE-TOOL BUILDERS' ATTACHMENTS
- ACCESSORY EQUIPMENT AND TOOLING
- ABILITY TO HANDLE SYMMETRICAL PARTS
- ERROR-SENSING EQUIPMENT

PROGRAM STOP CONTROL; RETURN TO ZERO

PICK UP PROGRAM AFTER STOP

AXIS NOMENCLATURE

TOOLING FEATURES

HOLDING FIXTURES – CLAMP SYSTEMS

ZERO POINTS AND CHECK POINTS; TOOL ORIENTATION

CUTTER SELECTION, SIZE, WEAR

CUTTER DYNAMICS, DEFLECTION, STRENGTH

CUTTER DESIGN (SPECIFICATIONS)

PART WARPAGE – TOOL DESIGN

PLATENIZED FIXTURES

TOOL-POSITION COORDINATION WITH MACHINE

USE OF INSTRUCTION SHEET (OPERATOR SET-UP DRAWING)

COMBINATION OF TOOLING FIXTURES

TYPE OF COOLANT

TOOL LIFE AND WEAR

INSPECTION TOOLS REQUIRED (SPECIAL DEVICES)

CUTTING-TOOL SUPPLY – INVENTORY CONTROL

PROGRAMMING

PRESENT PROGRAMMING METHODS

Manual

Computer-aided

Subroutine library

COMPUTER-CODED

LANGUAGE-CODED; "ASSEMBLY PROGRAM"

Automatic coding system

MANUAL

Basic training

REVIEW OF ANALYTIC GEOMETRY AND TRIGONOMETRY

VECTOR ANALYSIS

INTERPOLATION

BASIC DIFFERENTIAL CALCULUS

DESK CALCULATOR

ORIENTATION

Interdepartmental flow chart and responsibilities

TERMINOLOGY AND PROCEDURES FOR PROGRAMMING

Glossary of terms *such as set point, check point, coordinates, axes, and so forth*

Flow diagram; problem set up

SPECIAL FORMS AND PAPER WORK

Examples *such as manuscript, data sheet, set-up drawing, operator's instruction sheet*

EXAMPLE OF COMPLETE PROCESS

KEY-PUNCH TRAINING

COMPUTER AIDED

BASIC TRAINING

Review of analytic geometry and trigonometry

Vector analysis

ORIENTATION

Interdepartmental flow chart and responsibilities

TERMINOLOGY AND PROCEDURES FOR PROGRAMMING

Glossary of terms *such as set point, check point, coordinates, axes, computer language, coding language, and so forth*

Flow diagram; problem set-up

SPECIAL FORMS AND PAPER WORK

Examples *such as manuscript, data sheet, set-up drawing, operator's instruction sheet*

EXAMPLE OF COMPLETE PROCESS

METHODS OF CHECKING MANUSCRIPT

SUBROUTINE LIBRARY

BASIC TRAINING

Review of basic analytic geometry and trigonometry

ORIENTATION

Interdepartmental flow chart and responsibilities

TERMINOLOGY AND PROCEDURES FOR PROGRAMMING

Glossary of terms *such as set point, check point, coordinates, axes, computer language, coding language, and so forth*

CATALOG OF COMPUTER SUBROUTINES

CATALOG OF COMPUTER LANGUAGE CODE

FLOW DIAGRAM; PROBLEM SET-UP

SPECIAL FORMS AND PAPER WORK

Examples *such as manuscript, data sheet, set-up drawing, operator's instruction sheet*

EXAMPLE OF COMPLETE PROCESS

METHODS OF CHECKING MANUSCRIPT

AUTOMATIC PROGRAMMING

Since this method is presently in the development stage, it is felt that the principles and techniques being investigated should only be mentioned at this time.

PRINCIPLES OF DIGITAL COMPUTERS

The material outlined below is intended for trainees other than those associated with digital computer and programming groups.

BACKGROUND TECHNOLOGY

ABACUS

BABBAGE

FIRST RELAY MACHINE

FIRST ELECTRONIC MACHINE

OTHER MACHINES

CLASSES AND TYPES OF ELECTRONIC COMPUTERS AND AUXILIARY EQUIPMENT

CLASSES

Digital

Analog

TYPES

General purpose

Special purpose
example: directors

AUXILIARY

CAPABILITIES AND LIMITATIONS OF DIGITAL COMPUTERS

COMPARISON OF COMPUTER CHARACTERISTICS

Speed

Volume

Accuracy

Flexibility

Cost

HOW A COMPUTER WORKS

INPUT-OUTPUT SYSTEM

CONTROL SYSTEM

STORAGE SYSTEM

ARITHMETIC SYSTEM

LANGUAGE REQUIREMENTS FOR COMMUNICATION BETWEEN HUMAN AND COMPUTER

INTERNAL LANGUAGE

EXTERNAL LANGUAGE

ROLE OF THE COMPUTER IN NUMERICAL CONTROL

(describe with the aid of the following block diagram)

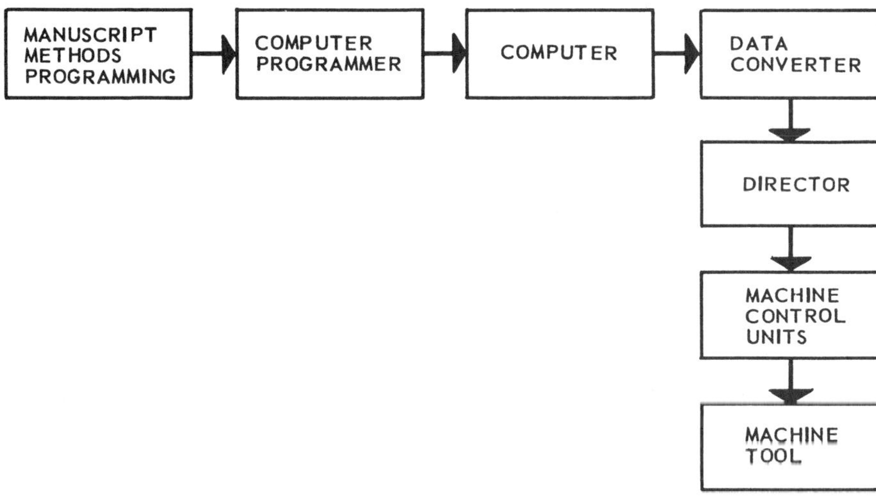

FUNCTIONS OF THE COMPUTER

Computations

Error checking

Coding

Program assembly

RELATION OF COMPUTER TO ADJOINING FUNCTIONS

FUTURE DEVELOPMENTS IN DATA PROCESSING FOR NUMERICAL CONTROL

GLOSSARY OF COMPUTER TERMS

APPENDIX E

The MIT–USAF–AIA Press Conference
Letter of Invitation, Program and List of Attendees

MASSACHUSETTS INSTITUTE OF TECHNOLOGY | CAMBRIDGE

Office of Public Relations

February 5, 1959

To the News Editor:

You are invited to attend a press conference on the APT System to be held at MIT from 9:30 AM until approximately 3:30 PM on Wednesday, February 25, to see a demonstration of a new system for automatically programming machine tools.

"APT" means Automatically Programmed Tools. The system is expected to result in savings of millions of dollars to the aircraft industry in coming years. It marks the "marriage" of digital computers and numerically controlled tools so that scientific methods can be applied to engineering and production methods.

The APT System was developed during the past two years through an unusual cooperative venture involving research by the MIT Servomechanisms Laboratory and field trial work by some 20 leading American aircraft plants. The system has the potential for being a major factor in the nation's industrial economy.

Working in cooperation with the Aircraft Industries Association and the Air Materiel Command, United States Air Force, we have prepared releases, pictures and other material on the system. This will be available in press kits at the conference, and advance releases will be mailed to you later this month. Representatives of all three agencies will present the background, history, technical details and potential of the system. A demonstration will be held at the MIT Computation Center and at the Servomechanisms Laboratory. There will be a period for questions and discussion.

Cameramen, both still and motion picture, will be welcome. All media representatives are invited to be our guests for cocktails and luncheon in the Campus Room of the Graduate House on the MIT campus. We have been fortunate in obtaining as a luncheon speaker Lt. General C. S. Irvine, Deputy Chief of Staff for Materiel, Hq. United States Air Force. Following the luncheon, another discussion of the APT System will be held for those who wish to remain for additional technical details.

You are asked to meet in the Little Theater of Kresge Auditorium on the West Campus, directly opposite the main entrance to MIT on Massachussets Avenue. We would appreciate all media representatives being present by 9:15 AM. Would you please return the enclosed card with the names of those planning to attend the conference so that we may receive it not later than Friday, February 20?

Sincerely,

Francis E. Wylie
Director of Public Relations

List of Conference Attendees

Ben Lee	Assist. Direc. Public Relations, AIA
James J. Haggerty	Missiles and Rockets Magazine
Robert Loebelson	Space Aeronautics
Allen M. Smythe	Magazine of Wall Street
Henry T. Simmons	Newsweek
Raymond H. Spiotta	Associate Editor, Machinery
Larry Tully	Aviation Week
Sam Cummins	Automotive Industry
R. W. Bolz	Editor, Automation
Byron K. Ledgerwood	Managing Editor, Control Engineering
William M. Stocker	American Machinist
Robert Whitlock	Electrical Engineering
Benedict Learburger	Product Engineering
Peter Kaprillyan	Technical Editor, Aircraft and Missiles
John R. Riggs	Associate Editor, Electrical Manufacturing
Harry Samuals	Popular Science
Glen Nixon	US News & World Report
Edward J. Egan, Jr.	Iron Age
Cal Campbell	Boston Herald, Popular Mechanics
Juan Cameron	Boston Herald
Reporter	Boston Daily Record
Tom Maguire	Electronics
Reporter	Boston Herald
Reporter	Boston Globe
David Muggeridge	Electronic News
Howard Rains	Electronic News
Alan Mayers	WGBH/FM
Barry Roach	The Tech
Jim Kistler	The Tech
Volta Torrey	Tech Review
Ruth Mehrtens	Time
David Wurzel	UPI Photos
Ed Fitzgerald	UPI Photos
Richard Growald	UPI Newsdesk
Brenton Willing, Jr.	Business Week
Loren Bullock	IBM
J. M. Wicker	IBM
J. P. Rankin	Convair, Fort Worth
Theodore Merrill	Business Week
Ralph Morse	WEEI (CBS)
Frank Prendergase	Industry
Albert Hughes	Christian Science Monitor
Photographer	Boston Globe
Harry Wharen	Metalworking
Roger May	Worcester Telegram
Arthur Gaskill	Movietone News
Wallace Dickson	Federal Reserve Bank (NE. Dis. Review)
Robert Cowen	Christian Science Monitor
Bob Ciroll	WBZ-TV
Gen. Irvine	USAF
Gen. Bergquist	USAF
Gen. McCormack	MIT
W. M. Webster	AMC

J. F. Reintjes	MIT
C. B. Perry	Douglas Aircraft
D. T. Ross	MIT
F. M. Verzuh	MIT Computation Center
O. Dale Smith	North American Aviation
D. F. Clements	MIT
Douglas Brown	MIT
Col. P. L. Hill	AMC
Major Hal Susskind	ADSID
Dean Stever	MIT
Edward Gleason	
Jeff Wylie	MIT
George Wood	MIT
F. L. Foster	DSR, MIT
Walter Milne	Press. Office, MIT

Press Conference
February 25, 1959

PROGRAM

9:30– 9:35	Welcome and opening remarks by Dr Gordon S. Brown, head of the Department of Electrical Engineering, MIT
9:35–10:40	Technical Session with talks by: J. A. Maurice – Aircraft Industries Association W. M. Webster – Air Materiel Command, USAF D. T. Ross – Servomechanisms Laboratory, MIT (detailed explanation of APT)
10:45–12:00	Demonstrations of APT on the IBM 704 Computer at the MIT Computation Center, and of numerically controlled machining at the Servomechanisms Laboratory.
12:05–12:30	Question period with panel of MIT, AMC, and AIA speakers, moderated by Prof J. F. Reintjes.
12:30– 2:30	Luncheon meeting, with address by Lt. General C. S. Irvine.
2:30– 3:30	Final question period.

INDEX